Physics

Revision Guide

Pauline Anning
Editor: Lawrie Ryan

Contents

Introduction

Key points

- At the start of each topic are the important points that you must remember.
- Anything marked with the Ⓗ icon is only relevant to those who are sitting the higher tier exams.

Synoptic link

Synoptic links show how the content of a topic links to other parts of the course. This will support you with the synoptic element of your assessment.

Study tip

Hints giving you advice on things you need to know and remember, and what to watch out for.

On the up

This feature suggests how you can work towards higher grades.

Using maths

This feature highlights and explains the key maths skills you need. There are also clear step-by-step worked examples.

This book has been written by subject experts to match the new 2016 specifications. It is packed full of features to help you prepare for your exams and achieve the very best that you can.

Key words are highlighted in the text. You can look them up in the glossary at the back of the book if you are not sure what they mean.

Many diagrams are as important for your understanding as the text, so make sure you revise them carefully.

Required practical

These practicals have important skills that you will need to be confident with for part of your assessment.

Higher

Anything in the higher tier spreads and boxes must be learnt by those sitting the higher tier exam. If you will be sitting foundation tier, you will not be assessed on this content.

In-text questions check your understanding as you work through each topic.

Summary questions

These questions will test you on what you have learnt throughout the whole chapter, helping you to work out what you have understood and where you need to go back and revise.

Practice questions

These questions are examples of the types of questions you may encounter in your exams, so you can get lots of practice during your course.

You can find brief answers to the summary questions and practice questions at the back of the book.

Checklists

Checklists at the end of each chapter allow you to record your progress, so you can mark topics you have revised thoroughly and those you need to look at again.

Chapter checklist

Tick when you have:

reviewed it after your lesson	✓		
revised once – some questions right			
revised twice – all questions right	✓	✓	

Move on to another topic when you have all three ticks

1 Energy and energy resources

Energy is needed to make objects move and to keep devices such as mobile phones working. Most of the energy you use is obtained by burning fuels, such as coal, oil, and gas.

The ability to access energy at the flick of a switch makes life much easier. People in developing countries often can't access energy as easily as we can, while people in developed countries are burning too much fuel and are endangering our planet by making the atmosphere warmer.

In this section you will learn about measuring and using energy. You will also learn how wind turbines and other energy resources that don't burn fuel could enable everyone to have access to energy.

I already know...	I will revise...
energy is a quantity that can be measured and calculated.	how to work out the energy stored in a moving object or in an object when it is lifted or stretched.
the total energy before and after a change has the same value.	how energy is stored and transferred and what happens to it after it is used.
energy transfers can be compared in terms of usefulness.	how to compare machines and appliances in terms of their efficiency.
energy transfer by heating can be reduced by using insulating materials.	how energy is transferred by heating through conduction.
energy is transferred by radiation.	how energy transfer by radiation is causing the Earth to become warmer.
the energy needed to heat an object depends on its mass and the material it is made of.	how to work out the energy needed to heat an object.
a renewable resource will not run out because it is a natural process.	how to compare different renewable and non-renewable energy resources.
burning fossil fuels releases carbon dioxide gas, which is a greenhouse gas, into the atmosphere.	how the environment is affected by the use of different energy resources.

Student Book pages 4–5 **P1**

1.1 Changes in energy stores

Key points

- Energy can be stored in a variety of different energy stores.
- Energy is transferred by heating, by waves, by an electric current, or by a force when it moves an object.
- When an object falls and gains speed, its gravitational potential energy store decreases and its kinetic energy store increases.
- When a falling object hits the ground without bouncing back, its kinetic energy store is transferred by heating to the thermal energy store of the object and surroundings, and by sound waves moving away from the point of impact.

Synoptic link

For more information on energy stores, look at Topics P2.1, P5.4, and P6.4.

Energy stores

- Energy can be stored in different ways.
  - Chemical energy stores include fuels, foods, and the chemicals found in batteries.
  - Kinetic energy stores describe the energy an object has because it is moving.
  - Gravitational potential energy stores describe the energy an object has because of its position relative to the ground.
  - Elastic potential energy stores describe the energy an object has because it is being stretched or squashed.
  - Thermal energy stores describe the energy an object has because of its temperature.
- Energy can be transferred from one store to another.
- Energy can be transferred by heating, waves, an electric current, or when a force moves an object.

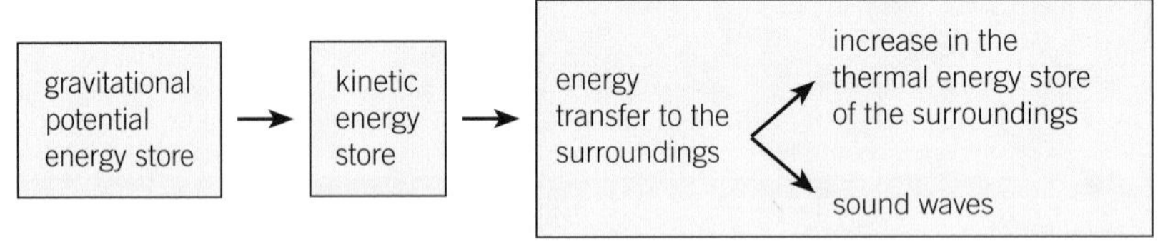

An energy transfer diagram for an object when it falls and when it hits the ground

1 What is the name of the energy store associated with a compressed spring?
2 What is the name of the energy store associated with a book on a high shelf?

Student Book pages 6–7 **P1**

1.2 Conservation of energy

Key points

- Energy cannot be created or destroyed.
- Conservation of energy applies to all energy changes.
- A closed system is an isolated system in which no energy transfers take place out of or into the energy stores of the system.
- Energy can be transferred between energy stores within

- Energy cannot be created or destroyed. This is called the principle of **conservation of energy** and it applies to all energy changes.
- Energy can be transferred usefully, stored, or dissipated.
- A **closed system** is an object or group of objects in which no energy transfers take place out of or into the energy stores of the system.
- However, as changes occur in a closed system, energy can be transferred to different stores within the closed system. For example:
  - When an object falls freely, the gravitational potential energy store of the object decreases and the kinetic energy store of the object increases.
  - When a person stretches an elastic band, the chemical energy store of the person's muscles decreases and the elastic potential energy store of the elastic band increases.

a closed system. The total energy of the system is always the same, both before and after any such transfers.

Synoptic link

For more information on energy, look at Topics P1.4, P1.5, and P1.6.

Study tip

Make sure you can explain the principle of conservation of energy.

Key words: conservation of energy, closed system

- When a pendulum swings towards the middle, the gravitational potential energy store of the pendulum decreases and its kinetic energy store increases. As the pendulum moves away from the middle, its kinetic energy store decreases and its gravitational potential energy store increases.

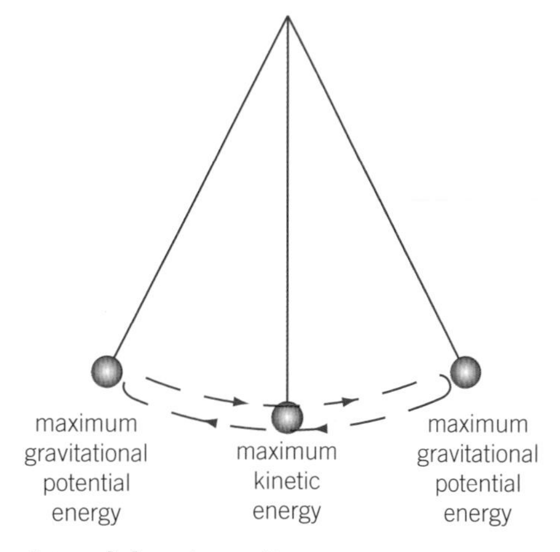

A pendulum in motion

1 What is meant by a closed system?
2 In terms of energy stores, what happens when you turn on a torch?

Student Book **pages 8–9** **P1**

1.3 Energy and work

Key points

- Work is done on an object when a force makes the object move.
- energy transferred = work done
- Work done is $W = F\,s$, where F is the force and s is the distance moved (along the line of action of the force).
- Work done to overcome friction is transferred as energy to the thermal energy stores of the objects that rub together and to the surroundings.

Synoptic link

For more information on work, look at Topics P1.4 and P5.4.

Key word: work

- To make a stationary object move, you need to apply a force to the object.
- When a force moves an object, energy is transferred to the object and **work** is done on it.
- When work is done to move an object, energy supplied to the object is equal to the work done in moving the object.
- Both work and energy have the unit joule, J.
- The work done on an object is calculated using the equation $W = F\,s$, where:

  W is the work done in joules, J

  F is the force in newtons, N

  s is the distance moved in the direction of the force in metres, m.
- If an object does not move when a force is applied to it, no work is done on the object.
- Friction is the force that opposes the motion of two surfaces in contact with each other.
- Work done to overcome friction is mainly transferred to thermal energy stores by heating.

1 What is the S.I. unit of energy?
2 What is the work done on an object when a force of 600 N moves it a distance of 12 m?

Student Book
pages 10–11 P1

1.4 Gravitational potential energy stores

Key points

- The gravitational potential energy store of an object increases when it moves up and decreases when it moves down.
- The gravitational potential energy store of an object increases when it is lifted up because work is done on it to overcome the gravitational force.
- The gravitational field strength at the surface of the Moon is less than on the Earth.
- The change in the gravitational potential energy store of an object is $\Delta E_p = m\, g\, \Delta h$

Study tip

An object's gravitational potential energy store is relative to another point – usually the surface of the Earth. So you will calculate changes in gravitational potential energy when an object is moved up or down, relative to the surface of the Earth.

- Gravitational potential energy is energy associated with an object because of its position in the Earth's gravitational field.
- Whenever an object is moved upwards, the energy in its gravitational potential energy store increases. This increase is equal to the work done on the object by the lifting force.
- Whenever an object is moved downwards, the energy in its gravitational potential energy store decreases. This decrease is equal to the work done by the gravitational force acting on it.
- When an object is lifted or lowered, the change in its gravitational potential energy is $\Delta E = m \times g \times \Delta h$, where:

  ΔE is the change in the gravitational potential energy in joules, J

  m is the object's mass in kilograms, kg

  g is the gravitational field strength in newtons per kilogram, N/kg

  Δh is the change in height in metres, m.
- The gravitational field strength at the surface of the Earth is 9.8 N/kg. The gravitational field strength at the surface of the Moon is about $\frac{1}{6}$ (one sixth) of this value.

1 What happens to the gravitational potential energy store of an object when the object moves downwards?

2 What is the unit of gravitational field strength?

Student Book
pages 12–13 P1

1.5 Kinetic energy and elastic energy stores

Key points

- The energy in the kinetic energy store of a moving object depends on its mass and its speed.
- The kinetic energy store of an object is $E_k = \frac{1}{2} m\, v^2$
- Elastic potential energy is the energy stored in an elastic

Kinetic energy

- All moving objects have kinetic energy. The greater the mass and the speed of an object, the more kinetic energy it has.
- Kinetic energy can be calculated using the equation $E_k = \frac{1}{2} \times m \times v^2$, where:

  E_k is the kinetic energy in joules, J

  m is the mass of the object in kilograms, kg

  v is the speed of the object in metres per second, m/s.

object when work is done on the object.

- The elastic potential energy stored in a stretched spring is $E_e = \frac{1}{2} k e^2$, where e is the extension of the spring.

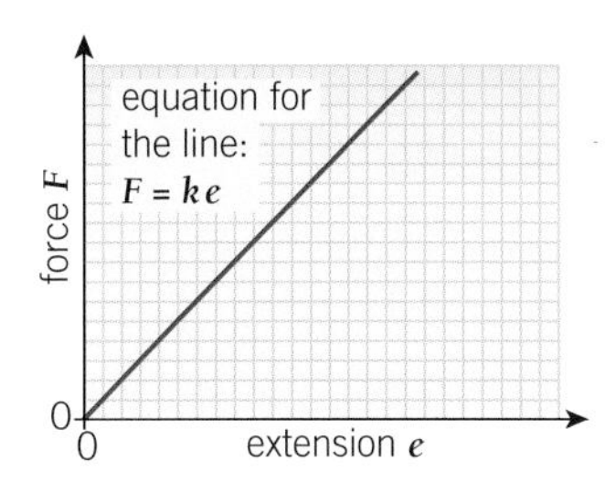

Force versus extension for a spring

On the up

To achieve the top grades, you should be able to rearrange the equations for kinetic energy and elastic potential energy and use them to perform calculations.

Elastic potential energy

Elastic potential energy is the energy stored in an elastic object that has been stretched or squashed.

- An object is described as being elastic if it regains its original shape after being stretched or squashed.
- When an elastic object is stretched or squashed, the work done on it is stored as elastic potential energy.
- When the object returns to its original shape, this energy becomes available for other transfers.
- The elastic potential energy of a spring can be calculated using the equation $E_e = \frac{1}{2} \times k \times e^2$ where:

  E_e is the elastic potential energy in joules, J

  k is the spring constant of the spring in newtons per metre, N/m

  e is the extension of the spring in metres, m.

1. What is the elastic potential energy of a spring with a spring constant of 150 N/m when its extension is 0.2 m?
2. What is the kinetic energy of a girl of mass 40 kg running at a steady speed of 5 m/s?

Student Book **pages 14–15** **P1**

1.6 Energy dissipation

Key points

- Useful energy is energy in the place we want it and in the form we need it.
- Wasted energy is the energy that is not useful energy and is transferred by an undesired pathway.
- Wasted energy is eventually transferred to the surroundings, which become warmer.
- As energy dissipates (spreads out) it gets less and less useful.

- Machines transfer energy for a purpose.
- **Useful energy** is energy transferred to where it is wanted in the form that is wanted. **Wasted energy** is energy that is not usefully transferred.
- Both the useful energy and the wasted energy will eventually be transferred to the surroundings, which will warm up. The energy is **dissipated**. As the energy spreads out it becomes more difficult to use for further energy transfers.
- Energy is often wasted because of friction between the moving parts of a machine. Sometimes friction may be useful – for example, in the brakes of a bicycle or a car. The kinetic energy store of the vehicle decreases and the thermal energy stores of the brakes and of the surroundings increase.

1. What happens to the wasted energy from a light bulb?
2. Why should we talk about energy being 'wasted' but not energy being 'lost'?

Key words: useful energy, wasted energy, dissipated

Student Book pages 16–17 P1

1.7 Energy and efficiency

Key points

- The efficiency of a device = useful energy transferred by the device ÷ total energy supplied to the device (× 100%)
- No energy transfer can be more than 100% efficient.
- Machines waste energy because of friction between their moving parts, air resistance, electrical resistance, and noise.
- (H) Machines can be made more efficient by reducing the energy they waste. For example, lubrication is used to reduce friction between moving parts.

Synoptic link

For more information on efficiency, look at Topics P1.8, P1.9, and P15.8.

Key word: efficiency

Study tips

As efficiency cannot be greater than 1 (or 100%), if you calculate an answer that is greater than this, go back and find the error in your working.

The only device that can be 100% efficient is an electric heater, which usefully transfers all of the energy supplied to it by heating its surroundings.

- The energy supplied to a machine (device) is called the input energy.
- The energy that is transferred usefully by the machine is called the useful output energy.
- From the principle of conservation of energy:

  input energy = useful output energy + energy wasted

- The less energy that is wasted by a machine, the more efficient the machine is.
- The **efficiency** of a device can be calculated using the equation

$$\text{efficiency} = \frac{\text{useful energy transferred by the device}}{\text{total energy supplied to the device}} \times 100$$

  Efficiency is a ratio, so it does not have a unit.

- No device can have an efficiency that is greater than 1 (or 100%).
- Different machines waste energy in different ways. These include friction between moving parts, air resistance, and electrical resistance.

1 In a machine, for every 25 J of energy supplied to the machine, 10 J are usefully transferred. Calculate the efficiency of the machine.

Higher

Machines can be made more efficient by reducing the energy they waste. For example: lubrication is used to reduce friction between moving parts; the shapes of moving vehicles are made more streamlined to reduce air resistance; and copper wires are used to minimise electrical resistance.

2 The shape of a lorry is modified to reduce air resistance. What happens to the efficiency of the lorry?

Efficiency

The efficiency of a device can be left as a decimal or can be multiplied by 100 to give a percentage. For example, an efficiency of 0.25 is the same as an efficiency of 25%.

On the up

You should be able to explain why a machine can never be more than 100% efficient. To achieve the top grades, you should be able to describe design features that can make a machine more efficient.

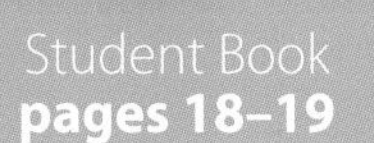

Student Book pages 18–19 **P1**

1.8 Electrical appliances

Key points

- Electricity, gas, and/or oil supply most of the energy you use in your home.
- Electrical appliances can transfer energy in the form of useful energy at the flick of a switch.
- Uses of everyday electrical appliances include heating, lighting, making objects move (using an electric motor), and producing sound and visual images.
- An electrical appliance is designed for a particular purpose and should waste as little energy as possible.

Energy in the home

Energy in the home is mostly supplied by electricity, gas, and oil. Electrical appliances are extremely useful because they transfer energy at the flick of a switch.

- Common electrical appliances include:

Appliance	Useful transfer
light bulb	light waves emitted from filament
electric mixer	work done by the blades of the mixer
speaker	sound waves from vibrations of speaker cone
television	light waves and sound waves

- Many electrical appliances transfer energy by heating. This may be a useful transfer, for example in a kettle, but for many other appliances, energy is wasted by heating.
- Appliances should be designed to waste as little energy as possible in order to make them as efficient as possible.

1 How is energy wasted by a light bulb?

2 a How is energy usefully transferred by a toaster?

b How is energy wasted by a toaster?

Student Book pages 20–21 **P1**

1.9 Energy and power

Key points

- Power is rate of transfer of energy.
- The power of an appliance is $P = \frac{E}{t}$
- efficiency of an appliance $= \frac{\text{useful power out}}{\text{total power in}} (\times 100)$
- power wasted by an appliance = total power input − useful power output

Key word: power

Study tip

Practise calculating the efficiency of a device as both decimals and percentages.

- The **power** of an appliance is the rate at which it transfers energy.
- The unit of power is the watt, W. 1 watt is equal to the rate of transferring 1 joule of energy in 1 second (i.e., 1 W = 1 J/s).
- Often a watt is too small a unit to be useful, so power is given in kilowatts (kW). 1 kW = 1000 W.
- Power can be calculated using the equation $P = \frac{E}{t}$, where:

  P is the power in watts, W

  E is the energy transferred in joules, J

  t is the time taken for the transfer in seconds, s.
- power wasted = total power supplied to device – useful power out put from device
- The efficiency equation can be written in terms of power:

$$\text{efficiency} = \frac{\text{useful power output}}{\text{total power input}} (\times 100)$$

1 How many watts are equivalent to 2.2 kW?

2 An electric motor transfers 36 kJ of energy in 2 minutes. What is the power output of the motor?

1. Name the energy store of a compressed spring. [1 mark]
2. Describe the changes to energy stores that take place when a child lifts a box from the floor to a high shelf. [2 marks]
3. Calculate the work done when a force of 1500 N pulls a car through a distance of 60 m in the direction of the force. [1 mark]
4. A ball of mass 0.45 kg is thrown vertically upwards and reaches a height of 3.0 m. The gravitational field strength is 9.8 N/kg. Calculate the change in gravitational potential energy of the ball. [2 marks]
5. Describe the changes to energy stores that take place when a student stretches and releases a spring. [3 marks]
6. A driver pushes on a broken-down car with a horizontal force, but the car does not move. Explain why no work is being done on the car. [2 marks]
7. Calculate the kinetic energy of a toy car with mass 50 g that is moving at a speed of 8.0 cm per second. [3 marks]
8. A student pushes a crate at a steady speed with a horizontal force of 200 N. It takes the student 15 seconds to move the crate a distance of 12 m.
   Calculate the power developed by the student in pushing the crate. [2 marks]
9. Describe the changes to energy stores that take place when a bungee jumper jumps off a bridge and eventually comes to a stop at the end of the rope. [4 marks]
10. A person pushes a trolley up a slope with a constant force at a constant speed.
    Explain why the work done on the trolley was less than the work done by the person. [3 marks]
11. An electric motor is used to raise a load. When the load gains 28 J of gravitational potential energy, 42 J of energy is wasted by the motor. Calculate the percentage efficiency of the motor. [3 marks]
12. A spring is stretched so it extends by 0.080 m. The elastic potential energy of the stretched spring is 1.2 J. Calculate the spring constant of the spring. [3 marks]

Chapter checklist

Tick when you have:

reviewed it after your lesson	✓		
revised once – some questions right	✓	✓	
revised twice – all questions right	✓	✓	✓

Move on to another topic when you have all three ticks

1.1 Changes in energy stores			
1.2 Conservation of energy			
1.3 Energy and work			
1.4 Gravitational potential energy stores			
1.5 Kinetic energy and elastic energy stores			
1.6 Energy dissipation			
1.7 Energy and efficiency			
1.8 Electrical appliances			
1.9 Energy and power			

Student Book pages 24–25 **P2**

2.1 Energy transfer by conduction

Key points

- Metals are the best conductors of thermal energy.
- Non-metal materials, such as wool and fibreglass, are the best insulators of thermal energy.
- The higher the thermal conductivity of a material, the higher the rate of energy transfer through the material.
- A thick layer of insulating material will have a low rate of energy transfer through it.

Synoptic link

For more information on reducing the rate of energy transfers at home, look at Topic P2.5.

Insulating a loft

Study tip

Give some examples of insulators and how they are used.

- Thermal conductors are materials that allow thermal energy to move through them easily. All metals are good thermal conductors.
- Thermal insulators are materials that do not allow thermal energy to move through them easily. Materials such as fibreglass and wool are good thermal insulators because they trap air.
- The rate of thermal energy transfer through a material depends on:
  - the temperature difference across the material
  - the thickness of the material
  - the thermal conductivity of the material.
- To reduce the rate of thermal energy transfer, the insulating material used should:
  - have as low a thermal conductivity as possible
  - be as thick as possible.

1 Why are saucepans often made of metal with wooden handles?

2 Why do materials that trap air have low thermal conductivities?

Testing sheets of materials as insulators

Use sheets of different materials to insulate identical cans of hot water.

Using a thermometer, measure the temperature of the water at the start of the investigation and after a fixed time.

Calculate the temperature change of the water during the fixed time.

When testing each insulating material, make sure the volume of water, the starting temperature of the water, and the fixed time before measuring the water's temperature all remain the same.

The material that is the best insulator is the one with the smallest temperature change in the fixed time.

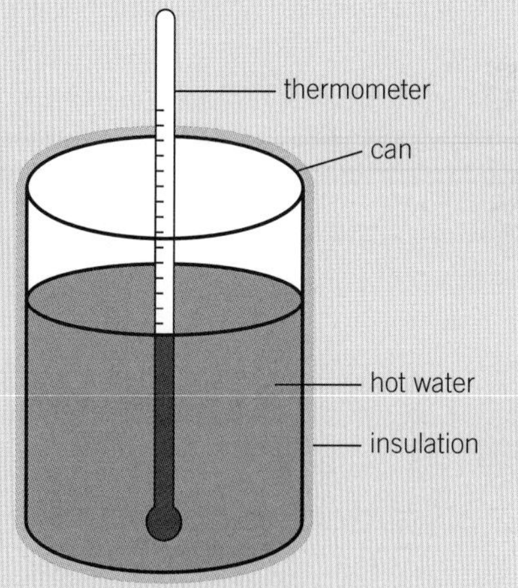

Student Book
pages 26–27

P2

2.2 Infrared radiation

Key points

- All objects emit and absorb infrared radiation.
- The hotter an object is, the more infrared radiation it emits in a given time.
- Black body radiation is radiation emitted by a perfect black body (a body that absorbs all the radiation that hits it).

Synoptic link

For more information on the electromagnetic spectrum, look at Topics P13.1 and P13.2.

Key words: infrared radiation, black body radiation

- Infrared waves are part of the electromagnetic spectrum. They are the waves with wavelengths a little longer than the wavelength of visible red light. We can detect **infrared radiation** with our skin – it makes us feel warm.
- All objects emit (give off) infrared radiation.
- The hotter an object is, the more infrared radiation it emits in a given time.
- An object at constant temperature emits infrared radiation at the same rate as it absorbs it.
- A perfect black body is an object that absorbs all the radiation that hits it.
- A perfect black body is also the best possible emitter of radiation. The radiation emitted by a perfect black body is called **black body radiation**.
- An object that has a constant temperature emits radiation across a continuous range of wavelengths.
- The intensity of the radiation an object emits has a peak at a certain wavelength. This peak wavelength depends on the temperature of the object.

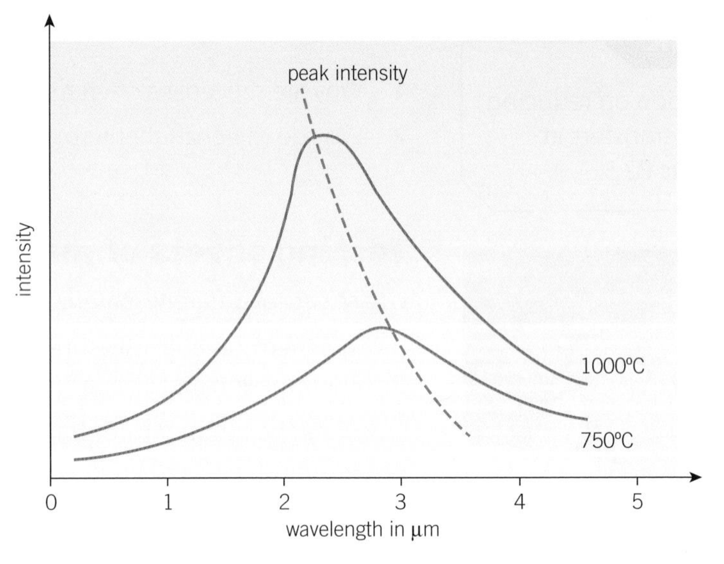

Black body radiation

1. What is infrared radiation?
2. How does the temperature of an object affect the rate at which it emits infrared radiation?
3. How does the intensity of black body radiation vary with wavelength?

~~2.3~~ More about infrared radiation

Key points

- The temperature of an object increases if it absorbs more radiation than it emits.
- The Earth's temperature is affected by the level of absorption of infrared radiation from the Sun, and the emission of infrared radiation from the Earth's surface and atmosphere.

Synoptic link

For more information on infrared radiation, look at Topics P2.2, P13.1, and P13.2.

Study tip

Draw a diagram to help you remember that surfaces that are good absorbers of infrared radiation are also good emitters of infrared radiation.

On the up

To achieve the top grades, you should be able to explain the factors that affect the temperature of the Earth.

- All objects absorb and emit infrared radiation.
  - An object at constant temperature absorbs and emits infrared radiation at the same rate.
  - If an object absorbs infrared radiation at a greater rate than it emits it, the temperature of the object increases.
  - If an object absorbs infrared radiation at a lesser rate than it emits it, the temperature of the object decreases.
  - An object with a light, shiny outer surface emits radiation at a lesser rate than an object with a dark, matt surface.
- The Earth receives light and infrared radiation from the Sun. Some of this radiation is reflected back into space, some is absorbed by the Earth's atmosphere, and some is absorbed by the Earth's surface.
- The temperature of the Earth depends on:
  - the rate at which radiation from the Sun is reflected or absorbed by the Earth's atmosphere, and the Earth's surface
  - the rate at which radiation is emitted from the Earth's surface, and from the Earth's atmosphere, into space.

1 Which surfaces are the best absorbers of infrared radiation?

2 Describe three things that may happen to infrared radiation from the Sun when it reaches the Earth's atmosphere.

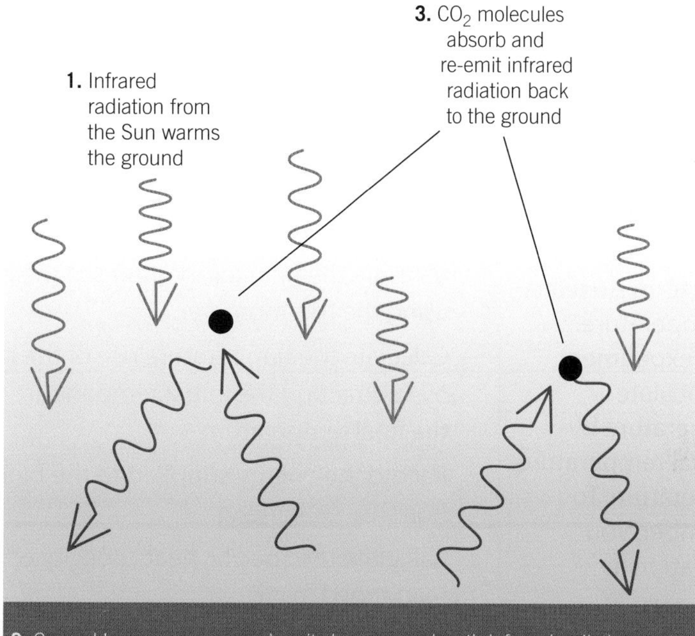

The absorption and emission of infrared radiation

Student Book
pages 30–31

P2

2.4 Specific heat capacity

Key points

- The specific heat capacity of a substance is the amount of energy needed to change the temperature of 1 kg of the substance by 1°C.
- Use the equation $\Delta E = mc\Delta\theta$ to calculate the energy needed to change the temperature of an object of mass m by temperature $\Delta\theta$.
- The greater the mass of an object, the more slowly its temperature increases when it is heated.
- To find the specific heat capacity, c, of a substance, use a joulemeter and a thermometer to measure ΔE and $\Delta\theta$ for a measured mass m of the substance. You can then calculate c using the equation $c = \frac{\Delta E}{m\Delta\theta}$

Key word: specific heat capacity

Study tips

Practise writing the equation for specific heat capacity, and stating what the terms mean.

Remember that the term $\Delta\theta$ refers to the *change* in temperature. Given data from an experiment, you may need to calculate this change in temperature by subtracting the initial temperature from the final temperature. To help yourself remember, you could write the equation as

$$\Delta E = m \times c \times (\theta_F - \theta_I)$$

where θ_I is the initial temperature and θ_F is the final temperature.

- When you heat a substance, you transfer energy to it. This increases its temperature. The **specific heat capacity** of the substance is the amount of energy required to raise the temperature of 1 kg of the substance by 1 °C (degree Celsius).
- The greater the specific heat capacity of a substance, the more energy you would need to supply to the substance in order to increase its temperature by 1 °C.
- For a given substance, a greater mass requires more energy to increase its temperature by 1 °C.

  You can calculate the energy required to change the temperature of a substance by a set amount, $\Delta\theta$, using the equation $\Delta E = m \times c \times \Delta\theta$, where:

  ΔE is the energy needed in joules, J

  m is the mass of the sample in kilograms, kg

  c is the specific heat capacity of the substance in joules per kilogram degrees Celsius, J/kg°C

  $\Delta\theta$ is the temperature change in degrees Celsius, °C.

1 The specific heat capacity of water is 4200 J/kg °C. How much energy is needed to raise the temperature of 2 kg of water by 1 °C?

2 The specific heat capacity of copper is 490 J/kg °C. How much energy is needed to raise the temperature of 1 kg of copper by 1 °C?

Measuring specific heat capacity

Measure the mass of the metal block using a top-pan balance.

Set up the apparatus as shown opposite.

Measure the initial temperature of the block using the thermometer.

Turn on the heater. Heat the metal block until the temperature has risen by about 10 °C.

Measure the final temperature of the block using the thermometer.

Calculate the temperature rise of the block by subtracting the initial temperature from the final temperature.

Record the energy supplied to the block from the joulemeter.

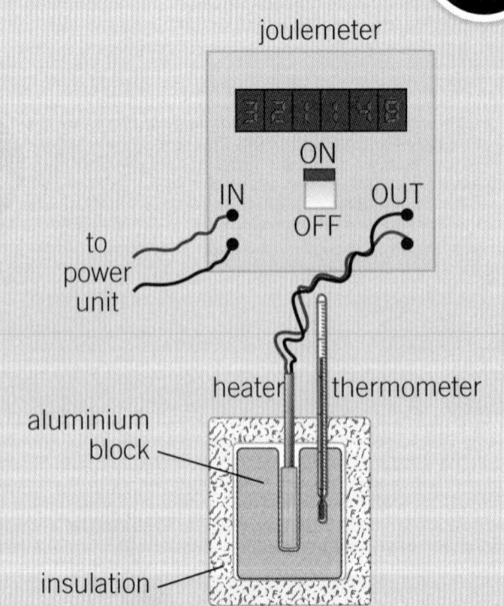

Measuring the specific heat capacity of a metal

Calculate the specific heat capacity of the metal using the equation in the rearranged form:

$$c = \frac{\Delta E}{m \times \Delta\theta}$$

This method can be used to measure the specific heat capacity of different metals, or of a liquid in an insulated beaker.

Student Book
pages 32–33 P2

2.5 Heating and insulating buildings

Key points

- Houses are heated by electric or gas heaters, oil or gas central heating systems, or by solid fuel in stoves or fireplaces.
- The rate of energy transfer from houses can be reduced by using loft insulation, cavity wall insulation, double-glazed windows, aluminium foil behind radiators, and having external walls built with thicker bricks that have lower thermal conductivity.
- Cavity wall insulation is insulation used to fill the cavity between the two brick layers of an external house wall.

Synoptic link

For more information on heating and insulating buildings, see Topics P2.1, P2.2, and P 2.3.

Study tip

Give some examples of thermal insulation that are commonly used to minimise energy transfer from buildings.

On the up

You should be able to describe how the rate of cooling of a building is affected by the thickness and thermal conductivity of its walls.

Whichever way a house is heated, people want to minimise the rate of energy transfer from their homes in order to reduce their fuel bills.

- Fibreglass loft insulation reduces the rate of energy transfer through the roof.
- Cavity wall insulation reduces rate of energy transfer through the outer walls.
- Aluminium foil behind radiators reflects infrared radiation back into the room.
- Double glazing reduces the rate of energy transfer through the windows.
- Thick bricks with low thermal conductivity reduce the rate of energy transfer through the exterior walls.

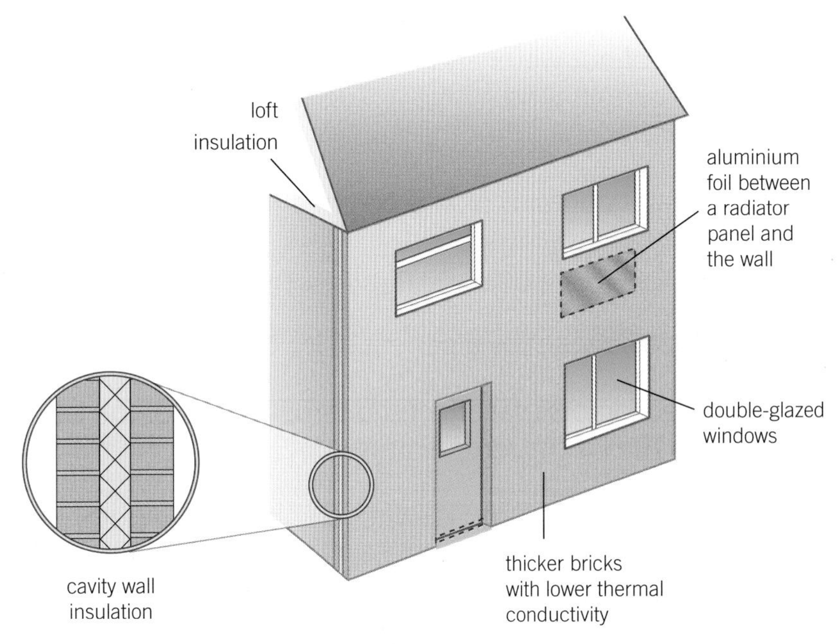

Reducing the rate of energy transfer

Solar panels absorb infrared radiation from the Sun.

- Solar cell panels generate electricity directly.
- Solar heating panels heat water directly.

1. Why are the pipes that contain the water in a solar heating panel often painted black?
2. Why are solar panels normally fitted on a roof facing south?

Summary questions

1. Give one example of a good thermal conductor. [1 mark]
2. Identify which surfaces are the best emitters of infrared radiation. [1 mark]
3. Name three things that affect the rate of transfer of thermal energy through a layer of material. [3 marks]
4. Explain why putting silver foil behind a radiator can reduce energy bills. [3 marks]
5. Explain the difference between a thermal conductor and a thermal insulator. [2 marks]
6. The specific heat capacity of copper is 490 J/kg °C.
   Calculate the energy needed to raise the temperature of 2.5 kg of copper by 3 °C. [2 marks]
7. A student investigates insulating materials by wrapping sheets of each material around identical cans of hot water and measuring the temperature decrease of the water in each can during a certain time.
   Suggest three variables that the student should control in this investigation. [3 marks]
8. A storage heater contains blocks that heat up overnight and then transfer energy to the room during the day.
   Explain why the blocks must be made from a material with a high specific heat capacity. [3 marks]
9. Describe what is meant by a perfect black body. [2 marks]
10. Describe three factors that affect the temperature of the Earth, and explain why each factor affects the Earth's temperature. [3 marks]
11. A low-voltage filament lamp is switched on and the potential difference across it is gradually increased.
    Explain why the lamp glows a dull red, followed by orange-red, and finally yellow-white. [3 marks]
12. An aluminium block of mass 1.5 kg was heated, and the following data were obtained:

    energy supplied to block = 16 200 J

    initial temperature of block = 20 °C

    final temperature of block = 32 °C

    Calculate the specific heat capacity of aluminium. [3 marks]

Chapter checklist

Tick when you have:

reviewed it after your lesson	✓	☐	☐
revised once – some questions right	✓	✓	☐
revised twice – all questions right	✓	✓	✓

Move on to another topic when you have all three ticks

2.1 Energy transfer by conduction	☐	☐	☐
2.2 Infrared radiation	☐	☐	☐
2.3 More about infrared radiation	☐	☐	☐
2.4 Specific heat capacity	☐	☐	☐
2.5 Heating and insulating buildings	☐	☐	☐

Student Book pages 36–37

P3

3.1 Energy demands

Key points

- Most of the energy you use comes from burning fossil fuels – mostly coal, oil, or gas.
- Nuclear power, biofuels, and renewable resources are also used to generate some of the energy you use.
- Uranium or plutonium is used as the fuel in a nuclear power station. Much more energy is released, per kilogram, from uranium or plutonium than from burning fossil fuels.
- Biofuels are renewable sources of energy. Biofuels, such as methane and ethanol, can be used to generate electricity.

Key words: biofuel, renewable, carbon-neutral, nuclear fuel, nucleus, reactor core

On the up

To achieve the top grades, you should be able to interpret information about energy resources from pie charts and graphs.

- In most power stations, water is heated to produce steam. The steam drives a turbine, which is coupled to an electrical generator that generates electricity.
- Most power stations in the UK burn fossil fuels, such as oil, coal, or gas, to generate electricity. Fossil fuels are obtained from long-dead biological material.
- A **biofuel** is any fuel obtained from living or recently living organisms.
  - Some biofuels can be used in small-scale, gas-fired power stations.
  - Biofuels are **renewable** sources of energy – they can be replaced at the same rate at which they are used up.
  - Biofuels are **carbon-neutral** – the carbon taken in as carbon dioxide from the atmosphere by the living organism can balance the amount released when the biofuel is burnt.
  - Waste vegetable oil, methane, rapeseed plants, ethanol, straw, nutshell, and woodchip are all examples of biofuels.

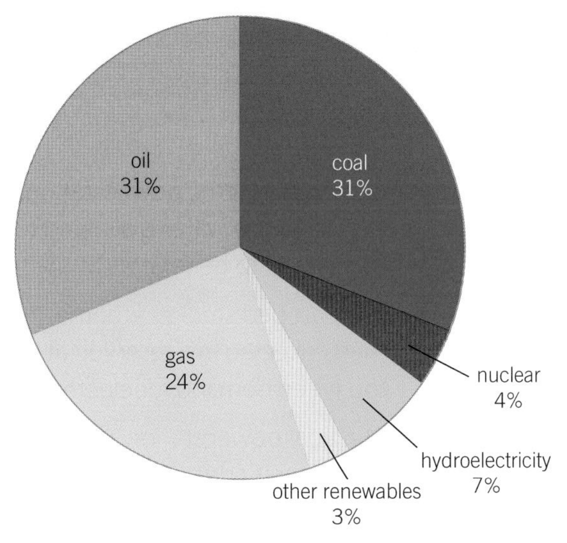

World energy demand and sources of energy in 2013

- The **nuclear fuel** used in a nuclear power station is uranium (or plutonium).
- The **nucleus** of a uranium atom is unstable and can split in two. This process releases energy.
- There are lots of uranium nuclei in the **reactor core**, so lots of energy can be released. The energy is used to heat water, turning it into steam.

Remember that the energy in a nuclear power station comes from a nuclear process. The uranium is not burnt.

1. What is meant by a non-renewable energy source?
2. Name three fossil fuels.
3. Using the information shown in the pie chart above, what percentage of world energy sources in 2013 were non-renewable?

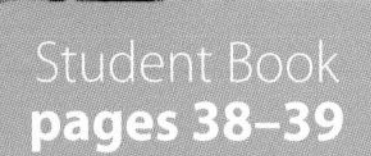

3.2 Energy from wind and water

Key points

- A wind turbine is an electricity generator, at the top of a narrow tower, whose turbine blades are turned by the wind.
- A wave generator uses the motion of waves to make a floating turbine move up and down.
- Hydroelectric turbine generators are turned by water running downhill.
- A tidal power station traps water from each high tide and uses it to turn turbines.

1. Which energy store is associated with the water in the reservoir of a hydroelectric power station?
2. Why is wind power not reliable?

- Wind, waves, and the tides are sources of renewable energy.
- In a wind turbine, the wind passing over the turbine blades at the top of a tall tower makes the turbine rotate and drive a generator.
- The movement of waves on the sea can be used to drive a floating turbine that turns a generator. The electricity is delivered to the grid system on shore by a cable.

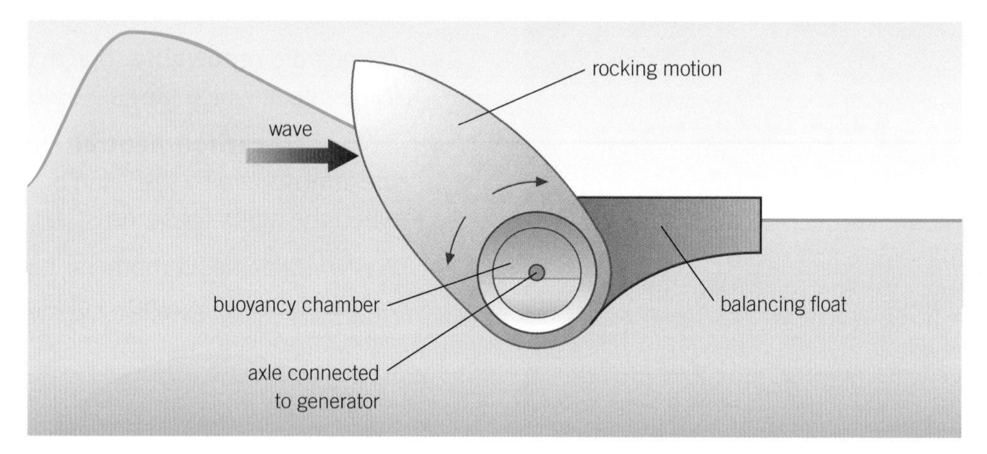

A wave generator

- At a hydroelectric power station, water is collected in a reservoir at the top of a hill. When the water is allowed to flow back downhill, it turns turbines that are connected to generators at the bottom of the hill.
- In a pumped storage scheme, the water at a hydroelectric power station is then collected in a reservoir at the bottom of the hill and pumped back to the top when demand for electricity is low, so the process can be repeated.
- At a tidal power station, water from each high tide is trapped behind a barrage. The water is released back into the sea through turbines. The turbines drive generators in the barrage.

A wind farm

A hydroelectric power station

Student Book pages 40–41 P3

3.3 Power from the Sun and the Earth

Key points

- Solar cells are flat, solid cells that use the Sun's energy to generate electricity directly.
- Solar heating panels use the Sun's energy to heat water directly.
- Geothermal energy comes from the energy released by radioactive substances deep within the Earth.
- In a geothermal power station, water is pumped onto hot rocks underground to produce steam, which drives turbines at the Earth's surface.

Key word: geothermal energy

Synoptic links

For more information on energy sources, look at Topics P3.2 and P3.4.

For more information on heating water, look at Topic 2.4.

Study tip

Name some situations where solar energy is particularly useful for generating electricity.

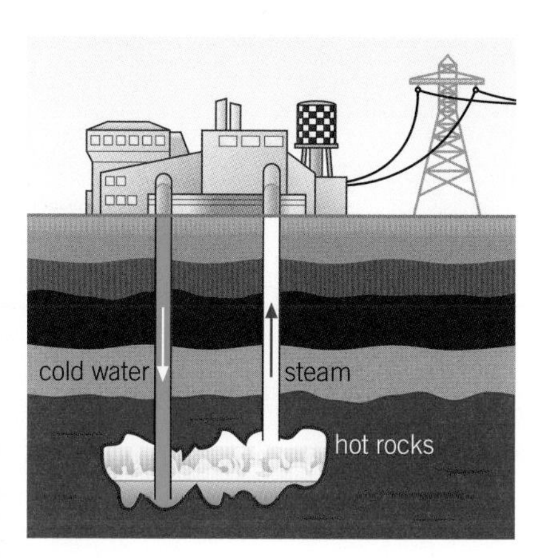

A geothermal power station

- Electromagnetic radiation from the Sun transfers energy to the Earth.
- A solar cell transfers this energy directly into electrical energy. Each cell only produces a small amount of electricity, so they are useful to power small devices, such as watches and calculators.
- Large numbers of solar cells can be joined together to form a solar panel. These can be useful to generate electricity in remote locations where there is no access to a national grid.
- A solar heating panel uses energy from the Sun to heat water that flows through the panel.

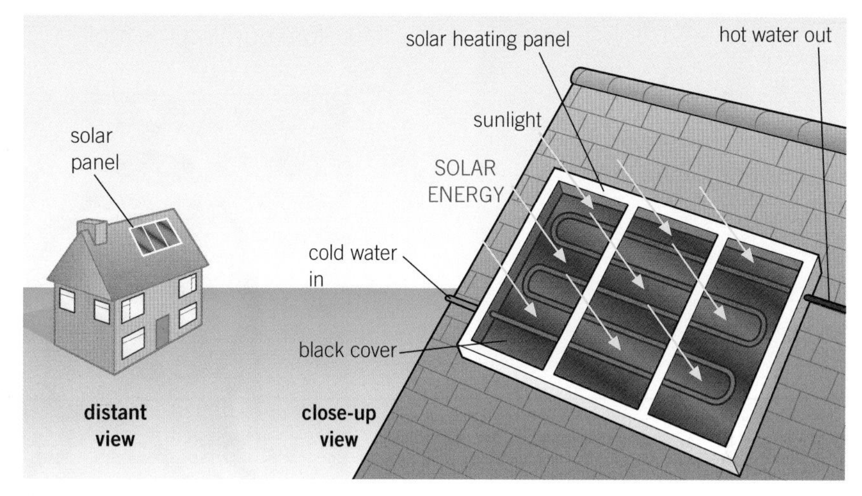

Solar water heating

- A solar power tower uses thousands of mirrors to reflect sunlight onto a water tank, which heats the water and produces steam. The steam is used to drive turbines that turn generators, and so produce electricity.
- **Geothermal energy** comes from energy released by radioactive substances deep within the Earth in volcanic or other suitable locations. This energy heats the surrounding rocks. In a geothermal power station, very deep holes are drilled and cold water is pumped down to the hot rocks. There it is heated and comes back to the surface as steam, which is used to drive turbines.

1 Why is a single solar cell only suitable to power a device such as a watch or calculator?

2 Why are only a few places in the world able to have geothermal power stations?

3.4 Energy and the environment

Key points

- When burnt, fossil fuels release greenhouse gases, which could cause global warming.
- Used nuclear fuels contain radioactive waste.
- Renewable energy resources will never run out, they do not produce harmful waste products (e.g., greenhouse gases or radioactive waste), and they can be used in remote places. However, they are not always reliable, they can cover large areas, and they can disturb natural habitats.
- Different energy resources can be evaluated in terms of reliability, environmental effects, pollution, and waste.

GAS OIL COAL

increasing greenhouse gas emissions

Greenhouse gases from fossil fuels

- Despite the fact that we rely so heavily on fossil fuels, such as coal, oil, and gas, they do have some disadvantages:
  - Fossil fuels are non-renewable energy resources.
  - Oil and gas will probably run out in the next 50 years or so, although coal will last much longer.
  - When coal, oil, or gas is burnt, greenhouse gases, such as carbon dioxide, are released. Most scientists believe this is causing global warming.
  - Burning coal and oil releases sulfur dioxide, which causes acid rain.
  - To deal with this waste, scientists are investigating ways to reduce the environmental impact of using fossil fuels, such as capturing and storing carbon dioxide in old oil and gas fields and removing sulfur from a fuel before burning the fuel.

The effects of acid rain

There are advantages and disadvantages to alternative sources of energy, too.

- Nuclear power – advantages
  - Nuclear power stations do not produce greenhouse gases.
  - Nuclear fuel transfers much more energy per kilogram than fossil fuels.
- Nuclear power – disadvantages
  - Used fuel rods contain radioactive waste that must be stored safely for centuries.
  - Nuclear power stations are safe in normal use, but an accident can make the surrounding area unsafe for many years.

- Renewable energy sources – advantages
  - They will not run out. They can be produced as fast as they are used.
  - They do not produce greenhouse gases or other dangerous waste.
  - They can be used where connection to the grid is uneconomical.
- Renewable energy sources – disadvantages
  - Renewable resources would not currently be able to meet world demand for energy.
  - Most renewable resources are not available all the time or can be unreliable.
  - Wind, tidal, and hydroelectric schemes can be an eyesore and can affect plant and animal life.
  - Solar cells need to cover large areas to generate large amounts of power.

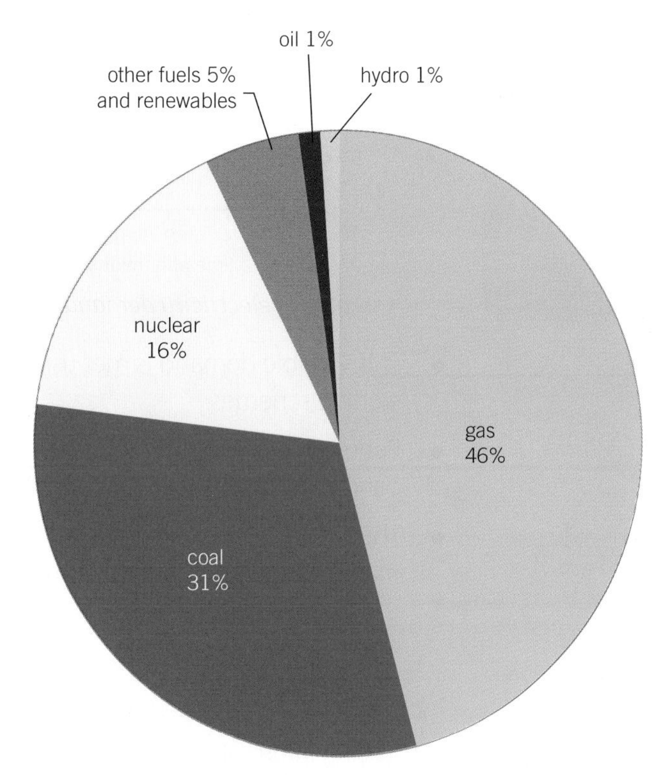

Energy sources for electricity

1. What type of area would be most suitable for a wind farm?
2. Why is wave power not reliable?
3. Using the pie chart opposite, what is the total percentage of electricity produced form burning fossil fuels?

Study tip

When discussing energy resources, remember that they all have both advantages and disadvantages. Make a table to show the advantages and disadvantages of the energy resources covered in this chapter.

Student Book pages 44–45 P3

3.5 Big energy issues

Key points

- Gas-fired power stations and pumped-storage stations can meet variations in electricity demand.
- Nuclear power stations are expensive to build, run, and decommission. Carbon capture of fossil fuel emissions is likely to be very expensive. Renewable resources are cheap to run but expensive to install.
- Nuclear power stations, fossil-fuel power stations that use carbon capture and storage, and renewable energy resources are all likely to be needed to supply energy in the future.

- Different types of power station have different start-up times.
- A constant amount of electricity is provided by nuclear and coal-fired power stations. This is called the base load demand.
- The demand for electricity varies at different times during the day, and between summer and winter.

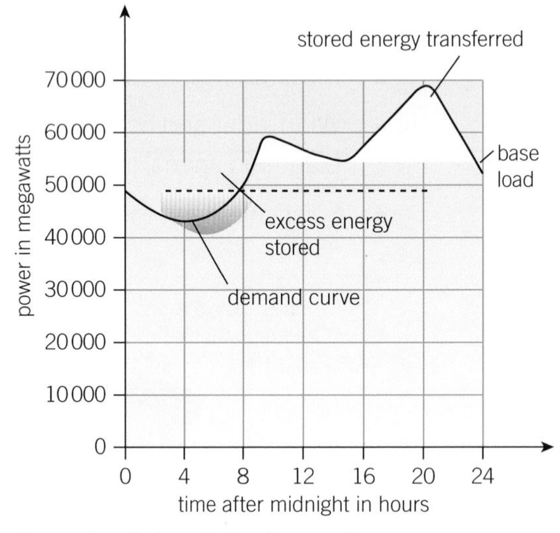

Example of electricity demand

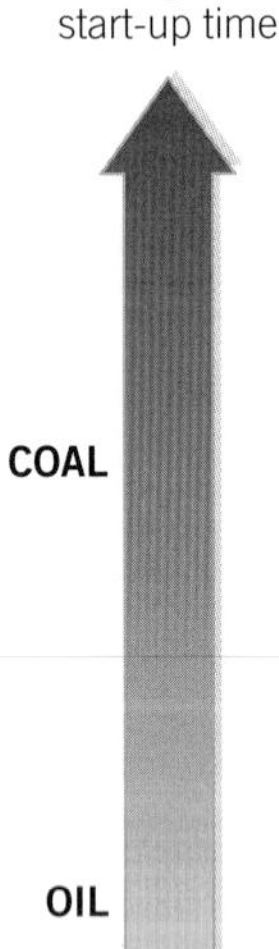

Start-up times of different types of power station

- This variable demand is met using gas-fired power stations and pumped storage schemes.
- Renewable energy resources are used when demand is high and the conditions for the resource are suitable.
- The overall cost of a new power station depends on building costs, fuel costs, maintenance costs, and decommissioning costs.

1 Why does the demand for electricity vary between summer and winter?
2 Why does the demand for electricity vary between day and night?
3 Using information from the graph above, at what time of day is the demand for electricity least and at what time is it greatest?

Study tip

Make a table to show the problems associated with the reliability of different renewable energy resources.

On the up

To achieve the highest grades, you should be able to evaluate the use of different energy resources in different situations.

1. Describe what is meant by a renewable resource. [1 mark]
2. Describe what a wind turbine is. [1 mark]
3. Name the types of power station used to meet the base load demand. [2 marks]
4. Describe how a hydroelectric power station works. [2 marks]
5. Explain why solar energy is not a reliable resource. [2 marks]
6. Give three examples of biofuels. [3 marks]
7. Describe two environmental problems caused by coal-fired power stations. [4 marks]
8. Describe the type of area needed to build a tidal power station. [3 marks]
9. Explain how a geothermal power station works. [4 marks]
10. A road sign on a remote country road is powered by a small wind turbine and a solar panel.
    Explain why a wind turbine and a solar panel are used together to power the sign. [3 marks]
11. A wind turbine produces an average power output of 2 MW in one 24-hour period.
    Calculate the energy output of the turbine during this time, in kWh. [3 marks]
12. Explain why a pumped storage power station is suitable for meeting sudden surges in demand. [3 marks]

Chapter checklist

Tick when you have:

reviewed it after your lesson	✓		
revised once – some questions right	✓	✓	
revised twice – all questions right	✓	✓	✓

Move on to another topic when you have all three ticks

3.1 Energy demands			
3.2 Energy from wind and water			
3.3 Power from the Sun and the Earth			
3.4 Energy and the environment			
3.5 Big energy issues			

01 A student heats some water in an electric kettle. When the kettle is switched on, energy is usefully transferred from the heating element in the kettle to the water in the kettle.

01.1 Name the process by which energy is transferred from the heating element to the water. [1 mark]

01.2 Describe how some energy is wasted while the kettle is switched on. [2 marks]

01.3 The student heated 0.300 kg of water. The change in temperature of the water was 75 °C. The specific heat capacity of water is 4200 J/kg °C. Calculate the energy transferred to the water. Choose the correct unit from the list below. [3 marks]

J **J/kg** **J/°C**

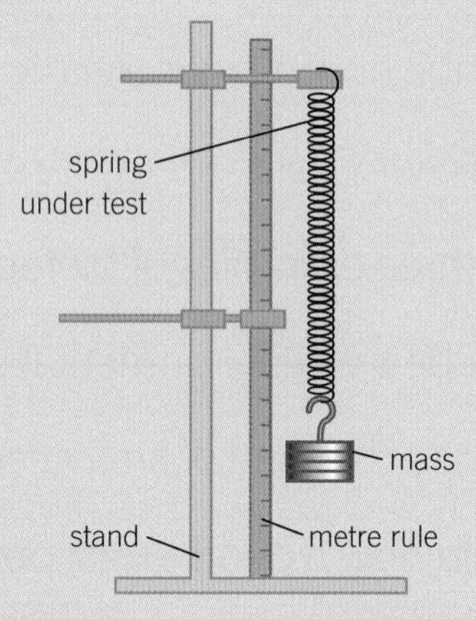

Figure 1

02 A student is investigating a spring. He hangs the spring from a clamp stand and attaches a mass to the bottom of the spring as shown in **Figure 1**.
He pulls the mass down a short distance and lets it go. The spring oscillates up and down. The student uses a stopwatch to time how long it takes for the spring to complete ten oscillations.

02.1 Complete the sentences to describe the energy transfers as the spring oscillates. Use answers from the box. You may use each answer once or not at all.

elastic potential **gravitational potential**
kinetic **thermal** **chemical**

When the mass is hanging from the stationary spring, the mass has a store of ______________ energy. When the student pulls the mass down, the spring stores energy as ______________ energy. As the student lets go, the mass begins to move upwards and its store of ______________ energy increases. [3 marks]

02.2 The student pulls the mass down by 0.025 m. The spring constant of the spring is 24 N/m. Calculate the extra energy now stored in the stretched spring. [2 marks]

02.3 Before carrying out the investigation, the student completed a risk assessment. To prevent the apparatus toppling over onto his feet, he put a heavy mass on the base of the clamp stand. Explain another safety precaution that should be taken when using the spring in this investigation. [3 marks]

Study tip

In a 'complete the sentence' question, read carefully through all of the sentences first, before you decide which words to use.

03 **Figure 2** shows the percentage of energy transferred from different parts of a house to the surroundings.

03.1 Calculate the percentage of the energy transferred through the walls. [1 mark]

03.2 On one day, 290 000 J of energy are transferred from the house to the surroundings. Calculate the energy transferred through the windows. [1 mark]

03.3 Fitting double-glazing would reduce the rate of energy transfer through the windows. Suggest one other advantage to the householder of double-glazed windows. [1 mark]

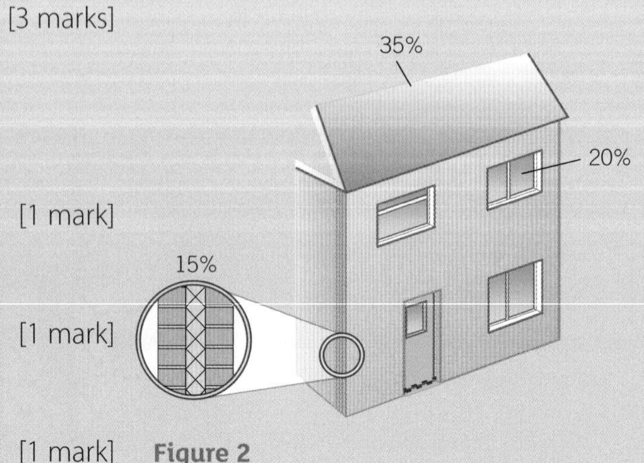

Figure 2

03.4 The house has walls with a cavity between them. The householder has the cavity filled with plastic foam containing bubbles of air. Explain how filling the cavity with plastic foam will affect the rate of cooling of the house. [4 marks]

04 **Figure 3** shows a solar power station. Sunlight falling on the curved mirrors is used to heat water.

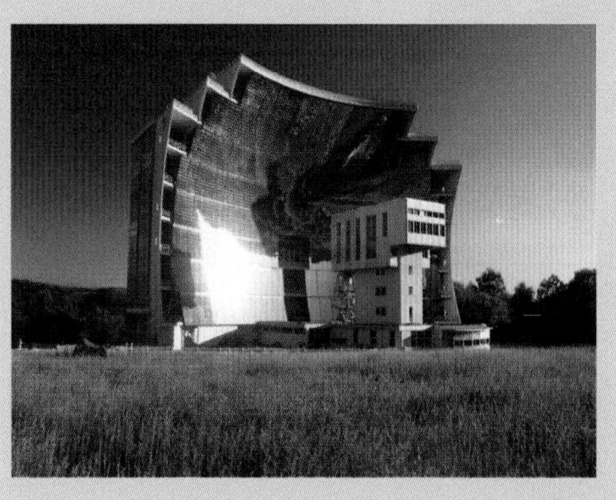

Figure 3

04.1 Energy sources can be renewable or non-renewable. Sunlight is a renewable energy source.
Name **two** other renewable energy sources. [2 marks]

04.2 Explain why the arrangement of mirrors rotates slowly during the day. [2 marks]

04.3 Describe the advantages and disadvantages of a solar power station compared with a coal-fired power station. [4 marks]

Study tip

Make sure that your answer includes both advantages and disadvantages and references to both types of power station.

05 A hairdryer contains an electric fan and a heating element.

05.1 Give two ways in which energy is usefully transferred by the heating element and the fan. [2 marks]

05.2 Give the main ways the hairdryer is likely to waste energy. [2 marks]

05.3 The hairdryer motor has an efficiency of 85%. The power supplied to the motor is 1600 W. Calculate the useful energy output of the motor when the hairdryer is used for 3.0 minutes. [4 marks]

06 A student wants a box that can be used to keep cans of drink cold on a sunny beach. She carries out an investigation using three identical boxes made from different materials. She puts cans of cold drink into each box and puts the lids on. She places a thermometer through a small hole in the lid of each box.
The student records the time it takes for the reading on each thermometer to increase by 5 °C.

06.1 Give the independent variable in this investigation. [1 mark]

06.2 Give the dependent variable in this investigation. [1 mark]

06.3 All the boxes are the same size and shape. Give **two** other control variables the student should use in this investigation. [2 marks]

06.4 The reading on the thermometer in the box made from expanded polystyrene took the greatest time to increase by 5 °C. Explain what conclusion can be made about the expanded polystyrene. [2 marks]

06.5 The student wraps the expanded polystyrene box in shiny metal foil. Explain how this will help to keep the cans colder for longer on a sunny beach. [2 marks]

2 Particles at work

All substances are made of atoms. Most atoms are stable and remain stable. Without this, the world as we know it wouldn't exist, and neither would we.

Every atom contains a nucleus surrounded by tiny particles called electrons. Atoms can lose or gain electrons, with different results. For example:

- Materials have different properties when the electrons in their atoms are shared in different ways.
- Metals conduct electricity because they contain electrons that have broken away from atoms inside the metal.
- Radioactive substances are made of atoms with unstable nuclei that emit harmful radiation when they become stable.

In this section you will learn about atoms as you learn about materials, electricity, and radioactivity.

I already know...	I will revise...
there are two types of electric charge.	how to calculate the charge flow in an electric circuit.
potential difference is measured in volts and current is measured in amperes.	how to work out the resistance and potential difference in an electric circuit.
a cell or battery pushes electrons round a circuit.	how mains electricity differs from electricity supplied by batteries.
power is how much energy is transferred per second.	how to calculate the power of an electrical appliance.
mass is the amount of matter in a substance and is measured in kilograms.	what we mean by density and how we can measure it.
gas particles move about very quickly and collide with the surface of the gas container.	how to explain why the pressure of a gas increases when it is heated in a sealed container.
the nucleus of an atom is composed of protons and neutrons.	how an unstable nucleus changes when it becomes stable and why the radiation it gives out is harmful.
energy is released when hydrogen nuclei fuse together in the Sun.	what nuclear fission and fusion are.

Student Book pages 50–51

P4

4.1 Electrical charges and fields

Key points

- Some insulating materials become charged when rubbed together.
- Electrons are transferred when objects become charged:
  - A material that becomes positively charged has lost electrons.
  - A material that becomes negatively charged has gained electrons.
- When brought together, like charges repel whilst unlike charges attract.
- The force between two charged objects is a non-contact force.

Key words: protons, neutrons, ion, static electricity, electric field

Study tip

Draw a diagram to help you remember that when an object is charged by friction, only electrons can be transferred.

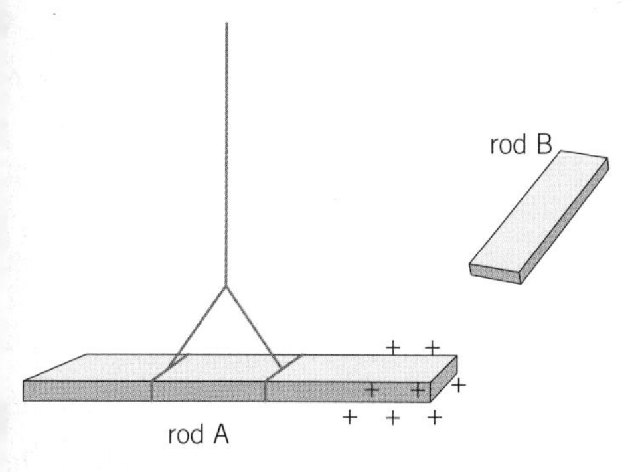

Investigating the force between charged objects. Rod A is suspended and rod B is brought close when it has either a negative or a positive charge

- The nucleus of an atom contains positively charged **protons** and neutral **neutrons**. Negatively charged electrons move around in the space outside the nucleus.
  - An uncharged atom has equal numbers of protons and electrons.
  - An atom may lose an electron to become a positively charged **ion**.
  - An atom may gain an electron to become a negatively charged ion.
- When you rub two electrically insulating materials together, electrons are rubbed off one material and deposited on the other. This is called **static electricity**. Which way the electrons are transferred depends on the particular materials.
- Electrons have a negative charge, so the material that has gained electrons becomes negatively charged. The one that has lost electrons is left with a positive charge. This process is called charging by friction.

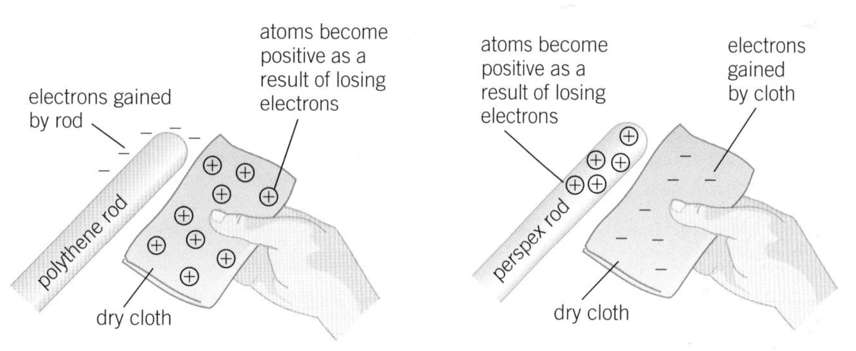

Charging by friction

- A charged object creates an **electric field** around itself. A second charged object in the field experiences a force.
  - The force acts at a distance, so is a non-contact force.
  - The bigger the distance between the objects, the weaker the force between them.
  - Two objects that have opposite electric charges attract each other.
  - Two objects that have the same electric charge repel each other.

1. How does an insulator become negatively charged?
2. What happens when two negatively charged objects are brought close together?

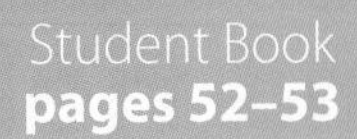

Student Book pages 52–53 **P4**

4.2 Current and charge

Key points

- Every component has its own agreed symbol. A circuit diagram shows how components are connected together.
- A battery consists of two or more cells connected together.
- The size of an electric current is the rate of flow of charge.
- The equation for the electric current of a circuit is $I = \frac{Q}{t}$
- The above equation can be rearranged to find charge flow or time: $Q = It$ or $t = \frac{Q}{I}$

Key word: electrons

- Every component in a circuit has an agreed circuit symbol. These are put together in a circuit diagram to show how the components in a circuit are connected together.
  - An electric current is a flow of charge. The charge is carried by **electrons**.
  - The size of an electric current is the rate of flow of electric charge and can be calculated using the equation $I = \frac{Q}{t}$, where:

    I is the current in amperes (or amps), A

    Q is the charge in coulombs, C

    t is the time in seconds, s.
- In a circuit that is a single closed loop, the current is the same at every point in the circuit.
- A diode allows current to flow in only one direction through the circuit.

1. What is the circuit symbol for a battery of three cells?
2. 10 C of charge pass a point in a circuit in 5 seconds. Calculate the current in the circuit.

Study tip

Draw a circuit diagram to help you remember that in an electric circuit, electrons flow from negative to positive, but we refer to 'conventional current' flow from positive to negative.

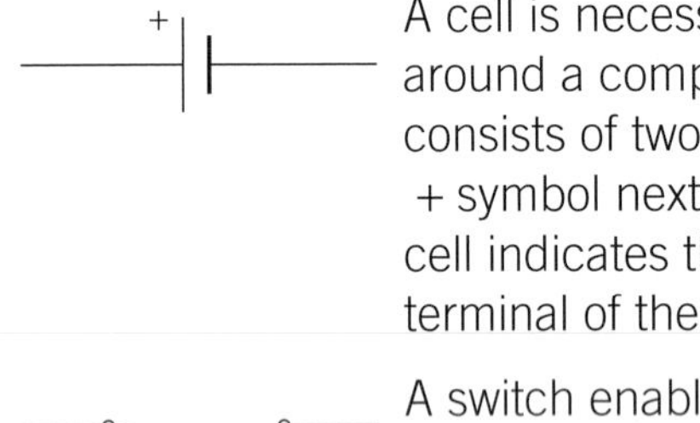

A cell is necessary to push electrons around a complete circuit. A battery consists of two or more cells. The + symbol next to the long line of the cell indicates that this is the positive terminal of the cell.

A switch enables the current in a circuit to be switched on or off.

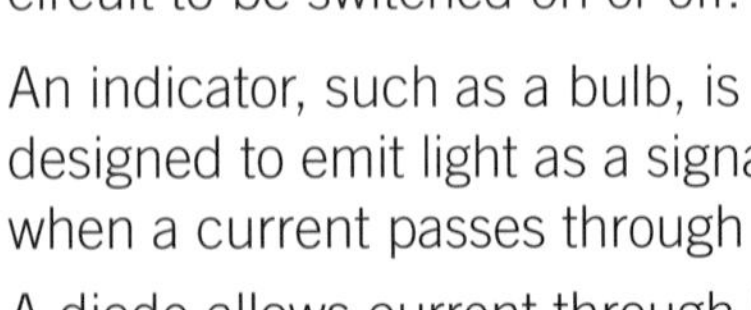

An indicator, such as a bulb, is designed to emit light as a signal when a current passes through it.

A diode allows current through in one direction only.

A light-emitting diode (LED) emits light when a current passes through it.

An ammeter is used to measure electric current.

A fixed resistor limits the current in a circuit.

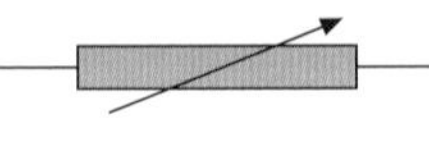

A variable resistor allows the current to be varied.

A fuse is designed to melt and therefore 'break' the circuit if the current through it is greater than a certain amount.

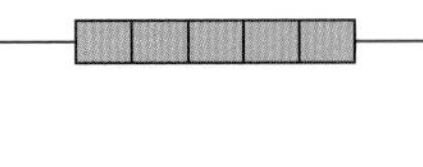

A heater is designed to transfer the energy from an electric current to heat the surroundings.

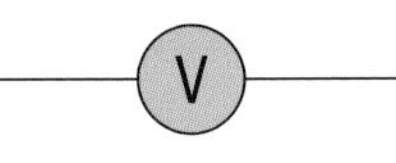

A voltmeter is used to measure potential difference (i.e. voltage).

Components and symbols

Student Book
pages 54–55

P4

4.3 Potential difference and resistance

Key points

- The equation for potential difference across a component is $V = \frac{\text{energy transferred } (E)}{\text{charge } (Q)}$
- The equation for resistance is $R = \frac{\text{potential difference } (V)}{\text{current } (I)}$
- Ohm's law states that the current through a resistor at constant temperature is directly proportional to the potential difference across the resistor.
- Reversing the potential difference across a resistor reverses the current through it.

Key words: series, potential difference, parallel, resistance

Study tip

Practise sketching the current–potential difference graph for a resistor at constant temperature.

1 Name the measuring instrument used to measure potential difference.

2 What is the unit of resistance?

- An ammeter measures the current passing through a component in a circuit. Ammeters are always connected in **series** with the component. The unit of current is the ampere (or amp), A.
- A voltmeter measures the **potential difference** (p.d.) across a component in a circuit. Voltmeters are always connected in **parallel** with the component. The unit of potential difference is the volt, V.
- Potential difference can be calculated using the equation $V = \frac{E}{Q}$, where:
  - V is the potential difference in volts, V
  - E is the energy transferred in joules, J
  - Q is the charge in coulombs, C.
- **Resistance** is the opposition to current flow.
- The resistance of a component can be calculated using the equation $R = \frac{V}{I}$, where:
  - R is the resistance in ohms, Ω
  - V is the potential difference in volts, V
  - I is the current in amperes, A.
- Current–potential difference graphs are used to show how the current through a component varies with the potential difference across it.

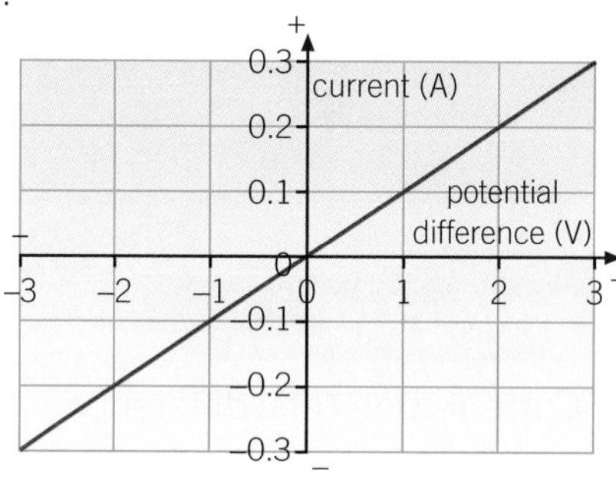

A current–potential difference graph for a resistor

  - If a resistor is kept at a constant temperature, its current–potential difference graph shows a straight line passing through the origin. This means the current is directly proportional to the potential difference across the resistor. This is called Ohm's law. Components that behave like this are called ohmic conductors.
  - Ohm's law only applies to a resistor at constant temperature
  - At constant temperature, a component's resistance remains constant, regardless of the direction of the potential difference and current.
  - The greater the resistance of the resistor, the less steep the current–potential difference graph.

How does the resistance of a wire depend on its length?

Set up the circuit shown.

Keep the current constant throughout the investigation by adjusting the variable resistor.

Measure the length of wire between the crocodile clips.

Measure the p.d. across the wire.

Calculate the resistance of the wire using $R = \frac{V}{I}$

Repeat with five different lengths of wire.

Plot a graph of length of wire (x-axis) against resistance of wire (y-axis).

Testing the resistance of a wire

Student Book pages 56–57 P4

4.4 Component characteristics

Key points

- The resistance of a component is $R = \frac{V}{I}$
- A filament lamp's resistance increases if the filament's temperature increases.
- Diode: forward resistance low; reverse resistance high.
- A thermistor's resistance decreases if its temperature increases.
- An LDR's resistance decreases if the light intensity on it increases.

Study tip

Current–potential difference graphs may be plotted with the current on the y-axis or on the x-axis. Try sketching the graphs shown on this page with the current on the x-axis to make sure you know the shape of the line either way round.

Key words: diode, light-emitting diode (LED), thermistor, light-dependent resistor (LDR)

1 What happens to the resistance of a thermistor if the temperature of the surroundings decreases?

2 What effect does reversing the p.d. across a filament lamp have on the current–potential difference graph for the lamp?

- The line on a current–potential difference graph for a filament lamp is a curve. So the current is **not** directly proportional to the potential difference.
- The resistance of the filament lamp increases as the current increases. This is because the resistance of the filament increases as its temperature increases.
- Reversing the potential difference through the filament lamp makes no difference to the shape of the curve. The resistance is the same for the same size current, regardless of the current's direction.

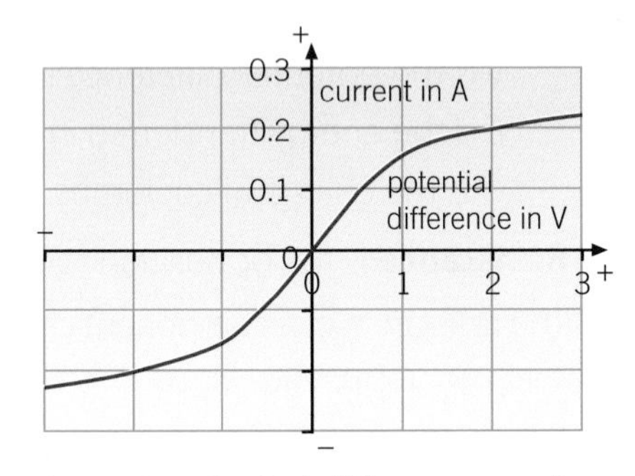

Current–potential difference graph for a filament lamp

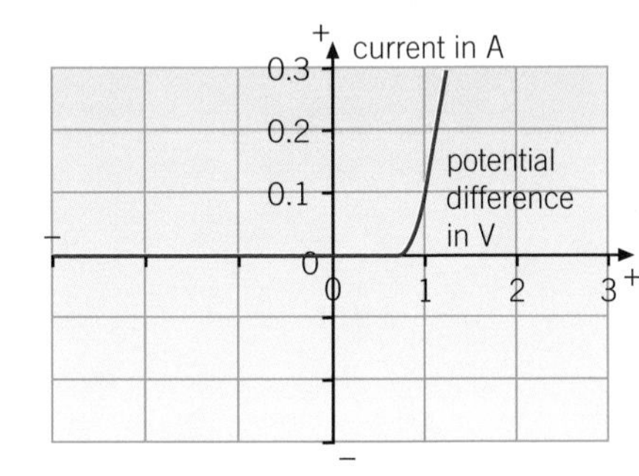

Current–potential difference graph for a diode

- The current through a **diode** is in one direction only. In the reverse direction the diode has a very high resistance so the current through it is virtually zero.
- A **light-emitting diode (LED)** emits light when a current passes through it in the forward direction.
- The resistance of a **thermistor** decreases as the temperature increases.
- The resistance of a **light-dependent resistor (LDR)** decreases as the light falling on it gets brighter.

Investigating different components

Set up the circuit shown.

Vary the current passing through the component by adjusting the variable resistor.

Record the current on the ammeter.

Record the p.d. on the voltmeter.

Calculate the resistance of the component using $R = \frac{V}{I}$

Repeat for five different values of current.

Reverse the connections to the battery and repeat the investigation.

Carry out this investigation for a resistor, a filament lamp, and a diode.

For each component, plot a graph of current against potential difference, including the negative section of the graph.

Investigating the resistance of different components

4.5 Series circuits

Key points

- For components in series:
  - the current is the same in each component
  - the total potential difference is shared between the components
  - adding their resistances gives the total resistance.
- For cells in series, acting in the same direction, the total potential difference is the sum of their individual potential differences.
- For resistors connected in series, total resistance $R_{total} = R_1 + R_2$
- Adding more resistors in series increases the total resistance because the current through the resistors is reduced and the total potential difference across them is unchanged.

On the up

To achieve the top grades, you need to be able to analyse series circuits to determine the current through and p.d. across different points in the circuit.

- In a series circuit the components are connected one after another. Therefore if there is a break anywhere in the circuit, charge stops flowing.
- There is no choice of route for the charge as it flows around the circuit, so the current through each component is the same.
- The current in a series circuit is determined by the potential difference (p.d.) of the power supply and the total resistance of the circuit, governed by the equation $I = \frac{V}{R}$
- The p.d. of the supply is shared between all the components in the circuit. So the p.d.s across individual components add up to give the p.d. of the supply.

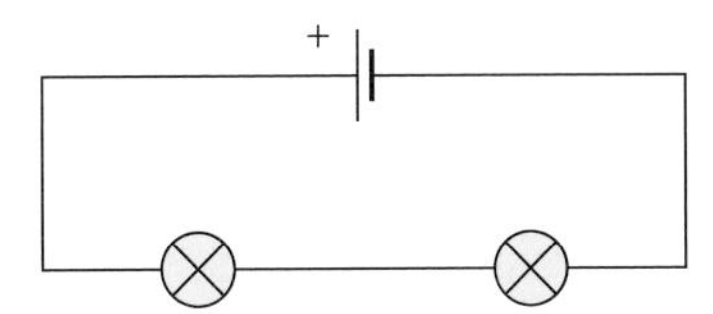

Bulbs in series

- The resistances of the individual components in series add up to give the total resistance of the circuit. $R_{total} = R_1 + R_2 + \ldots$
- The bigger the resistance of a component, the bigger its share of the p.d.

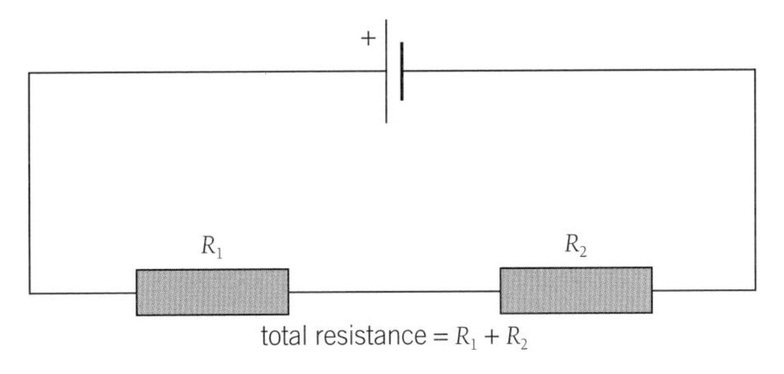

Resistors in series

1 What happens in a series circuit if one component stops working?

2 How could you calculate the total resistance in a series circuit?

Study tip

Draw a circuit diagram to help you remember that in a series circuit the current is the same through all components.

Student Book pages 60–61 P4

4.6 Parallel circuits

Key points

- For components in parallel:
  - the total current is the sum of the currents through the separate components
  - the potential difference across each component is the same.
- The bigger the resistance of a component, the smaller the current that will pass through that component.
- The current through a resistor in a parallel circuit is $I = \frac{V}{R}$
- Adding more resistors in parallel decreases the total resistance because the total current through the resistors is increased and the total potential difference across them is unchanged.

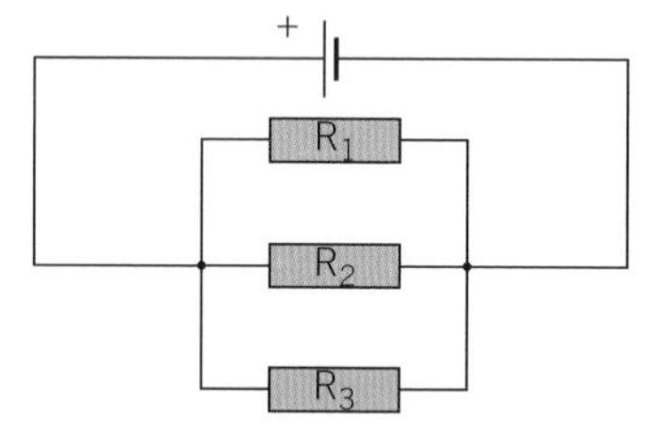

Resistors in parallel

Study tip

Explain how you could do a simple test on a line of 'fairy lights' to see if the bulbs are connected in series or parallel.

On the up

To achieve the top grades, you need to be able to analyse parallel circuits to determine the current at different points.

- In a parallel circuit each component is connected across the supply, so if there is a break in one branch of the circuit, charge can still flow in the other parts.
- Each component is connected across the supply potential difference (p.d.), so the p.d. across each component is the same.
- There are junctions in the circuit so different amounts of charge can flow through different components. The current through each component depends on the component's resistance. The bigger the resistance, the smaller the current through that component.
- The current through a component in a parallel circuit can be calculated using $I = \frac{V}{R}$
- The total current though the whole circuit is the sum of the currents through the separate branches.
- The total resistance of two (or more) components in parallel is less than the resistance of the component with the least resistance.
- Many circuits are a mixture of both parallel parts and series parts.

1 What happens in a parallel circuit if one component stops working?
2 In a parallel circuit, what is the relationship between the supply p.d. and the p.d. across each parallel branch?

Testing resistors in series and parallel

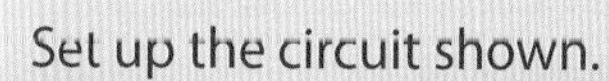

Set up the circuit shown.

For the first resistor, record the current on the ammeter and the p.d. on the voltmeter.

Calculate the resistance of the resistor using $R = \frac{V}{I}$

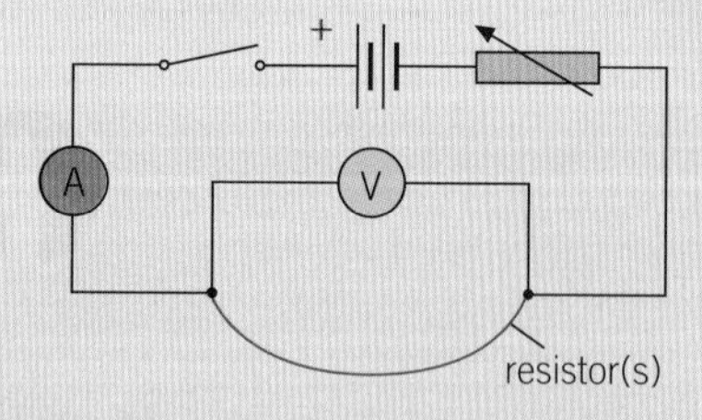

Testing resistors

Repeat with a second and third resistor and calculate their resistances.

Connect the resistors in the circuit in series with each other.

Repeat the measurements and use these to calculate the total resistance of the resistors in series. The total resistance of the resistors in series should equal the sum of their individual resistances.

Connect the resistors in the circuit in parallel with each other.

Repeat the measurements and use these to calculate the total resistance of the resistors in parallel. The resistance of the resistors in parallel should be less than the resistance of the resistor with the least resistance.

1. Write the type of charge that an electron has. [1 mark]
2. A 15 Ω resistor and an 18 Ω resistor are connected in series.
Calculate their total resistance. [1 mark]
3. A charge of 18 C passes through a lamp in 4.0 seconds.
Calculate the current passing through the lamp. [2 marks]
4. The current through a resistor is 4.0 A when the potential difference across it is 24 V.
Calculate the resistance of the resistor. [2 marks]
5. Explain, in terms of electrons, how an insulator becomes positively charged. [3 marks]
6. Describe how an ammeter and a voltmeter should be placed in a circuit to measure the current through, and the potential difference across, a resistor. [2 marks]
7. A series circuit contains a variable resistor. The resistance of the variable resistor is increased.
Explain what happens to the potential difference across the variable resistor. [2 marks]
8. Draw a circuit diagram to show a circuit containing a cell, a bulb, a resistor, and a switch all connected in series. [3 marks]
9. One branch of a parallel circuit contains a bulb and a variable resistor connected in series.
A second branch contains an LED and a resistor connected in series. The two branches are connected in parallel to each other and to a battery.
Draw a circuit diagram for the circuit. [5 marks]
10. Describe what is meant by an ohmic conductor. [2 marks]
11. The potential difference across a bulb is 4.5 V.
Calculate the energy transferred to the bulb as 8.0 C of charge passes through it. [2 marks]
12. Sketch and explain the shape of the line on a current–potential difference graph for a diode. [4 marks]

Chapter checklist

Tick when you have:

reviewed it after your lesson	✓		
revised once – some questions right	✓	✓	
revised twice – all questions right	✓	✓	✓

Move on to another topic when you have all three ticks

4.1 Electrical charges and fields			
4.2 Current and charge			
4.3 Potential difference and resistance			
4.4 Component characteristics			
4.5 Series circuits			
4.6 Parallel circuits			

Student Book
pages 64–65

P5

5.1 Alternating current

Key points

- Direct current (d.c.) flows in one direction only. Alternating current (a.c.) repeatedly reverses its direction of flow.
- A mains circuit has a live wire, which is alternately positive and negative every cycle, and a neutral wire at zero volts.
- The peak potential difference of an a.c. supply is the maximum voltage when measured from zero volts.
- To calculate the frequency of an a.c. supply, measure the time period of the waves and then use the equation
  $$\text{frequency} = \frac{1}{\text{time taken for 1 cycle}}$$

Synoptic links

For more information on mains electricity, look at Topic P5.5.

For more on the frequency of a wave, look at Topic P12.2.

For more information on transformers, look at Topics P15.7 and P15.8.

Key words: direct current, alternating current, live wire, neutral wire, National Grid, step-up transformer, step-down transformer

On the up

You should be able to explain the difference between direct current and alternating current.

- Cells and batteries supply current that passes round the circuit in one direction only. This is called **direct current** (d.c.).
- The current from the mains supply repeatedly reverses its direction. This is called **alternating current** (a.c.).
- Frequency is the rate at which an alternating current reverses its direction. The frequency of the UK mains supply is 50 hertz (Hz), which means that mains current reverses its direction of flow 50 times each second.
- If you measure the time taken for the current to complete one full cycle, you can calculate the frequency using the equation
  $$\text{frequency} = \frac{1}{\text{time taken for 1 cycle}}$$
- The **live wire** of the mains supply alternates between a positive and negative potential difference with respect to the **neutral wire**, which stays at zero volts.
- The live wire alternates between peak p.d.s of +325 V and –325 V (see graph). In terms of electrical power, this is equivalent to a direct current power supply with a p.d. of 230 V.

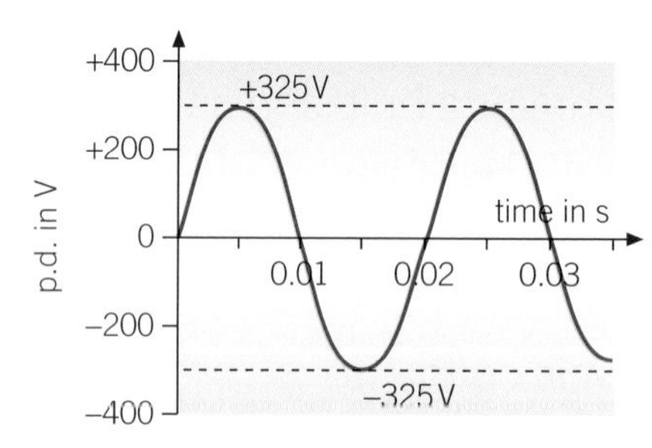

Mains p.d. against time

- The **National Grid** is a system of cables and transformers linking power stations to consumers.
  - **Step-up transformers** are used to increase the p.d. from the power station to the transmission cables.
  - **Step-down transformers** are used to decrease the p.d. to a much lower value for domestic use.

1 What is direct current?

2 What is the potential of the neutral wire?

Student Book pages 66–67

P5

5.2 Cables and plugs

Key points

- Sockets and plugs are made from stiff plastic materials that enclose the electrical connections. Plastic is used because it is a good electrical insulator.
- A mains cable is made up of two or three insulated copper wires surrounded by an outer layer of flexible plastic material.
- In a three-pin plug or a three-core cable, the live wire is brown, the neutral wire is blue, and the earth wire is striped green and yellow.
- The earth wire is connected to the longest pin in a plug and is used to earth the metal case of a mains appliance.

Mains cable

Key words: earth wire, fuse

- Most electrical appliances are connected to the sockets of the mains supply using a cable and a three-pin plug.

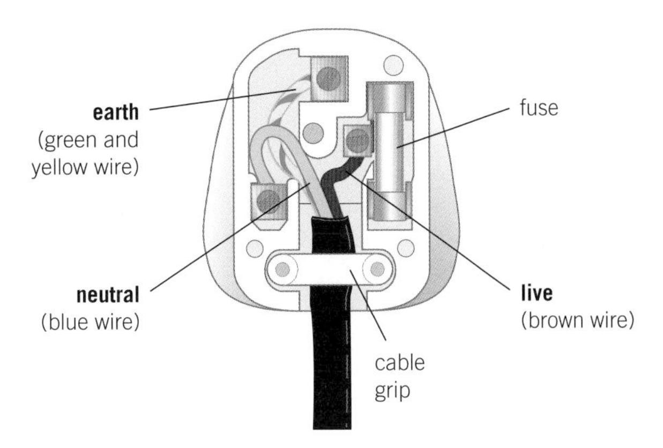

Inside a three-pin plug

- The outer cover of a three-pin plug and the outer casing of sockets are made from a stiff plastic material that is a good insulator.
- The pins of the plug are made of brass. Brass is a good electrical conductor. It is also hard and will not rust or oxidise.
- The green and yellow striped wire (of a three-core cable) is connected to the earth pin. A two-core cable doesn't have an **earth wire**.
- The brown wire is connected to the live pin and the blue wire is connected to the neutral pin.
- The outer coverings of the wires are made from a flexible plastic material.
- The plug contains a **fuse** between the live pin and the live wire. If the current in the fuse becomes too big, it will melt and cut off the current to the live wire.
- Appliances with a metal case must be earthed – the case is attached to the earth wire in the cable, which stops the metal case becoming live if the live wire breaks and touches the case.
- Appliances with plastic cases do not need to be earthed. They are said to be double-insulated, and are connected to the supply with a two-core cable containing just a live wire and a neutral wire.
- Cables of different thicknesses are used for different purposes. The larger the current to be carried, the thicker the cable needs to be.

1 Why is the outer covering of the wires made from a flexible plastic material?
2 Which is the longest pin in a plug?

Study tip

To help you remember the colours of the wires in a three-pin plug, practise sketching the inside of the plug using coloured pens of the correct colour to draw the wires.

5.3 Electrical power and potential difference

Key points

- The power supplied to a device is the energy transferred to it each second.
- The energy transferred to a device is $E = P \times t$
- The electrical power supplied to an appliance is equal to $P = I \times V$
- To determine the correct rating for a fuse, work out the normal current through the appliance using the equation $I = \frac{P}{V}$

Synoptic link

For more information on power, look at Topics P1.9 and P5.5.

Study tip

Take care with units in calculations. The power of an appliance is often given in kilowatts, kW, where 1 kW = 1000 W. Practise doing power calculations and converting the answers to kW.

- When electrical charge flows through an appliance, electrical energy is transferred, by the appliance, to another energy store. The rate at which energy is transferred is called the power.
- The power of an appliance can be calculated using the equation $P = \frac{E}{t}$, where:

  P is power in watts, W

  E is energy in joules, J

  t is time in seconds, s.
- This equation can be rearranged to give $E = P \times t$
- The power of an appliance can also be calculated using the equation $P = I \times V$, where:

  P is the power appliance in watts, W

  I is the current in amperes, A

  V is the potential difference in volts, V.
- Electrical appliances have their power rating shown on them. The p.d. of the mains supply is 230 V. The power equation can be rearranged to give $I = \frac{P}{V}$ and used to calculate the normal current through an appliance. Hence, you can work out the correct size of fuse for the appliance.
- The fuse is chosen so that its current rating is slightly higher than the current through the appliance.

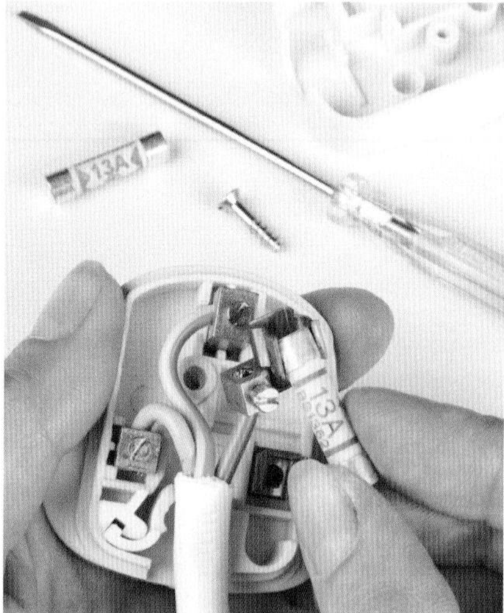

Changing a fuse

1. What is a fuse?
2. Calculate the power, in kW, of an appliance that transfers 90 000 J of energy in 30 seconds.

Student Book
pages 70–71

P5

5.4 Electrical currents and energy transfer

Key points

- The charge flow is $Q = I \times t$
- When charge flows through a resistor, energy transferred to the resistor makes it hot.
- The energy transferred to a component is $E = V \times Q$
- When charge flows around a circuit for a given time, the energy supplied by the battery is equal to the energy supplied to all the components in the circuit.

Synoptic link

For more information on the energy of electrical appliances, look at Topic P1.9.

- An electric current is a flow of charge.
- The charge that flows in a given time can be calculated using the equation $Q = I \times t$, where:

  Q is the charge in coulombs, C

  I is the current in amperes, A

  t is the time in seconds, s.
- When charge flows through a resistor, energy is transferred to the resistor. The thermal energy store of the resistor increases, so the resistor becomes hotter. This thermal energy is transferred to the surroundings.
- The energy transferred to a component by a flow of charge can be calculated using the equation $E = V \times Q$, where:

  E is the energy in joules, J

  V is the potential difference in volts, V

  Q is the charge in coulombs, C.
- Total energy transferred by the supply is equal to the sum of the energies transferred to each component in the circuit.

1. Calculate how much charge flows past a particular point in a circuit in 5 minutes when the current in the circuit is 2.0 A.
2. An appliance has a p.d. of 230 V across it. Calculate how much energy is transferred when a charge of 200 C flows through the appliance.

5.5 Appliances and efficiency

Key points

- A domestic electricity meter measures how much energy is supplied to a house.
- The energy supplied to an appliance is $E = P \times t$
- useful energy supplied by a device = efficiency of device × energy supplied to device

Synoptic link

For more information on energy and power, look at Topic P1.9.

Study tip

When doing efficiency calculations, check whether efficiency has been given as a decimal or a percentage. Practise converting between decimals and percentages.

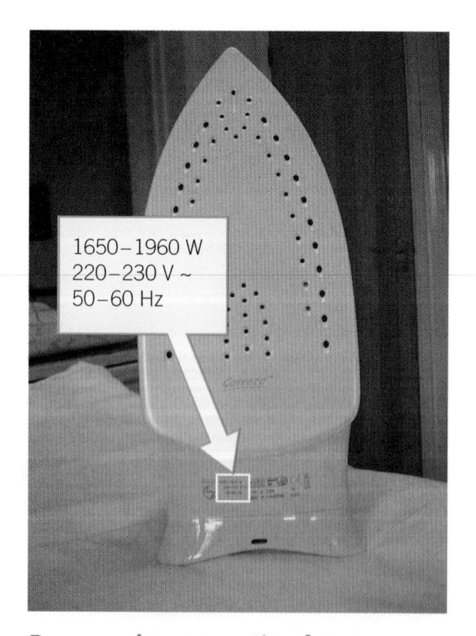

Power and energy rating for a household appliance

- Householders pay for the amount of electrical energy they use. This is measured by their domestic electricity meter.
- The energy transferred *to* an appliance by the mains supply can be calculated using the equation $E = P \times t$, where:

  E is the energy transferred to the device from the mains supply in joules, J

  P is the power of the electrical appliance in watts, W

  t is the time in seconds, s.
- The power of an appliance is given on a plate attached to the appliance.
- Electrical appliances waste energy because of the heating effect of the current in the wires. Appliances with electric motors also waste energy due to friction between the moving parts.
- The efficiency of an electrical appliance can be calculated using the equation

  $$\text{efficiency} = \frac{\text{useful energy transferred by appliance}}{\text{total energy supplied to appliance}} (\times 100) \text{ or}$$

  $$\text{efficiency} = \frac{\text{output power}}{\text{input power}} (\times 100)$$
- The useful energy transferred by the appliance is sometimes called the output energy.
- Efficiency may be expressed as a decimal value between 0 and 1, or as a percentage.

1. For every 25 J of energy that are supplied to a light bulb, only 5 J are usefully transferred. Calculate the efficiency of the light bulb.
2. A 2 kW kettle is switched on for 3 minutes. Calculate how much energy is supplied to the kettle in this time.

On the up

To achieve the top grades, you should be able to analyse data to determine the efficiency of different electrical appliances.

1. Give the frequency of the mains a.c. supply in the UK. [1 mark]
2. For a mains three-core cable, name the colours of the outer plastic covering on
   - **a** the live wire
   - **b** the neutral wire
   - **c** the earth wire in a mains. [3 marks]
3. Give the reason why appliances with plastic cases do not need to be earthed. [1 mark]
4. The current in a mains heater is 10 A.
   Calculate the power of the heater. [2 marks]
5. Describe what is meant by alternating current. [2 marks]
6. A resistor has a potential difference of 230 V across it.
   Calculate how much energy is transferred when a charge of 350 C flows through the resistor. [2 marks]
7. Explain why a two-core cable can be used for a hairdryer. [2 marks]
8. Explain why the pins of a mains plug are made from brass. [3 marks]
9. The frequency of the a.c. mains supply in the USA is 60 Hz.
   Calculate the time taken for one complete cycle of the USA mains supply. [2 marks]
10. A UK householder has 1 A, 3 A, and 13 A fuses available.
    Explain which fuse should be used in a 650 W table lamp that plugs in to the mains supply. [3 marks]
11. A hairdryer has an input power of 1800 W and an efficiency of 83%.
    Calculate the output power of the hairdryer. [2 marks]
12. The time taken for one complete cycle of the UK mains a.c. supply is 0.02 seconds, and the peak p.d. is 325 V.
    Sketch a graph of mains a.c. p.d. against time for two complete cycles, adding appropriate numbers to the axes. [3 marks]

Chapter checklist

Tick when you have:

reviewed it after your lesson	✓		
revised once – some questions right	✓	✓	
revised twice – all questions right	✓	✓	✓

Move on to another topic when you have all three ticks

5.1 Alternating current	☐	☐	☐
5.2 Cables and plugs	☐	☐	☐
5.3 Electrical power and potential difference	☐	☐	☐
5.4 Electrical currents and energy transfer	☐	☐	☐
5.5 Appliances and efficiency	☐	☐	☐

Student Book **pages 76–77** P6

6.1 Density

Key points

- density $= \frac{\text{mass}}{\text{volume}}$ (in kg/m^3)
- To measure the density of a solid object or a liquid, measure its mass and its volume, then use the density equation $\rho = \frac{m}{V}$
- Rearranging the density equation gives $m = \rho \times V$ or $V = \frac{m}{\rho}$
- Objects that have a lower density than water (i.e., $< 1000\ kg/m^3$) float in water.

Study tip

When doing a density calculation, write down the units for each quantity to be substituted and check that the units of your answer match these.

Synoptic links

For more information on density, look at Topic P11.2.

For more information on density and floating, look at Topic P11.4.

Key word: density

- The **density** of a substance is defined as its mass per unit volume.
- Density can be calculated using the equation $\rho = \frac{m}{V}$, where:

  ρ is density in kilograms per cubic metre, kg/m^3

  m is mass in kilograms, kg

  V is volume in cubic metres, m^3.
- Alternative units for density are g/cm^3 (where $1000\ kg/m^3 = 1\ g/cm^3$).
- Objects with a density that is less than the density of a given liquid will float in that liquid. So objects with a density less than $1000\ kg/m^3$ will float in water.

1 The SI unit for density is kg/m^3.
What is an alternative unit for density?

2 A block of aluminium has a mass of 2.0 kg and a volume of $7.4 \times 10^{-4}\ m^3$.
What is the density of aluminium in kg/m^3?

Density tests

For a solid object, measure the mass using an electronic balance.

For a *regular solid*, such as a cuboid, measure the lengths of the sides using a millimetre rule, vernier callipers, or a micrometer screw gauge. Use these measurements to calculate the volume.

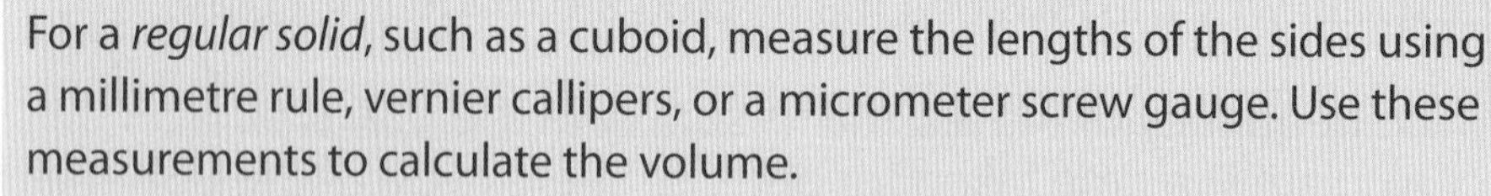

For an *irregular solid*, lower the object into a measuring cylinder of water until it is submerged. Calculate the volume from the rise in water level.

For a liquid, measure the mass of an empty beaker. Tip the liquid into the beaker and measure the new mass. Subtract the mass of the beaker to give the mass of the liquid.

Measure the volume of the liquid by pouring it into a measuring cylinder.

In each case, use the equation $\rho = \frac{m}{V}$ to calculate the density.

The volume of a cuboid $= a \times b \times c$

volume of cuboid $= a \times b \times c$

The volume of a cuboid

Student Book
pages 78–79

P6

6.2 States of matter

- The three states of matter are: solid, liquid, and gas.
- Substances can change from one state to another when they are heated or cooled.
- Changes of state are **physical changes** because no new substances are produced.
- When a substance changes state, the number of particles in the substance stays unchanged, so the mass of the substance is the same before and after the change of state.

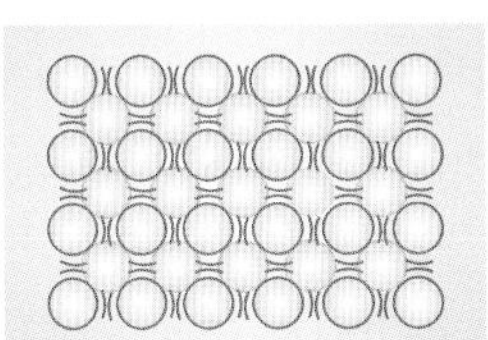

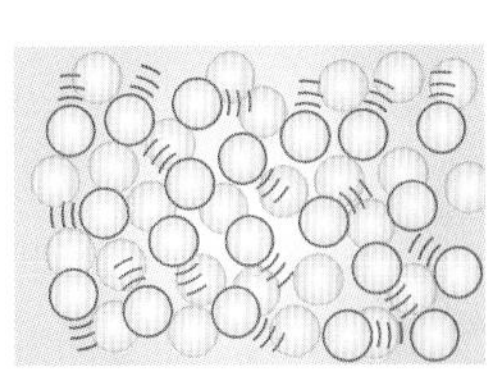

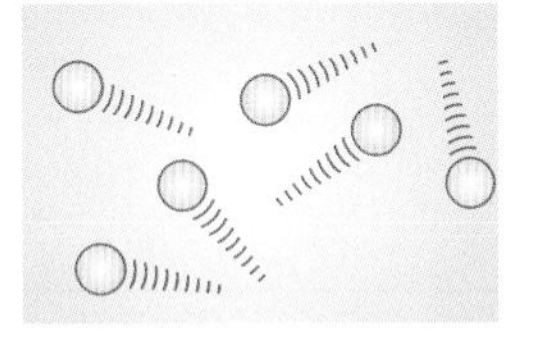

The arrangement of particles in solid state, liquid state, and gas state

- In a solid, the particles vibrate about fixed positions so the solid has a fixed shape.
- In a liquid, the particles are in contact with each other but can move about at random. Hence, a liquid doesn't have a fixed shape, and it can flow.
- In a gas, the particles are usually far apart and move about at random, much faster than the particles in a liquid. So the density of a gas is much less than that of a solid or liquid.
- The particles in a substance in its solid, liquid, or gas state have different amounts of energy depending on the state. For a given amount of a substance, its particles have the most energy in the gas state and the least energy in the solid state.

1 How is the arrangement of the particles in a liquid different from that in a solid?

2 How is the arrangement of the particles in a gas different from that in a liquid?

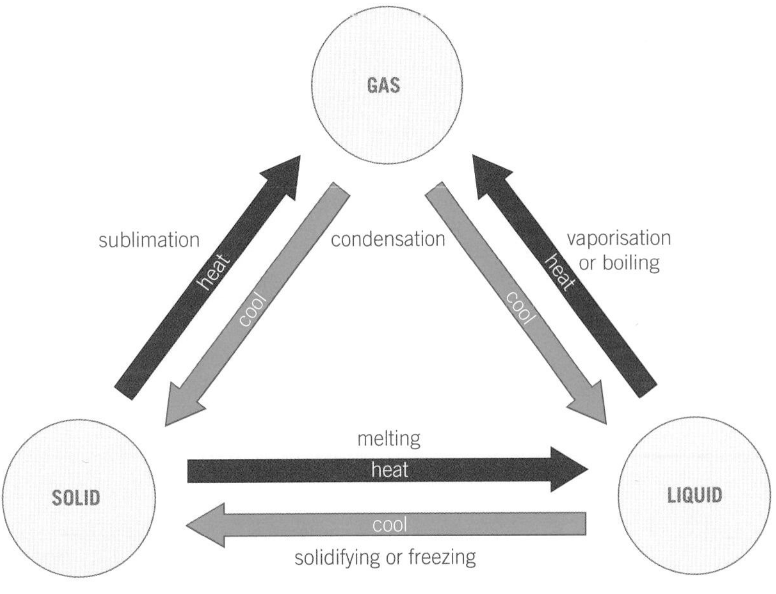

Changes of state

Key points

- The particles of a *solid* are held next to each other in fixed positions. They are the least energetic of the states of matter.
- The particles of a *liquid* move about at random and are in contact with each other. They are more energetic than particles in a solid.
- The particles of a *gas* move about randomly and are far apart (so gases are much less dense than solids and liquids). They are the most energetic of the states of matter.
- When a substance changes state, its mass stays the same because the number of particles stays the same.

Synoptic link

For more information on changes of state, look at Topics P6.3 and P6.5.

Key word: physical changes

Study tip

To help you learn about the three states of matter, practise drawing the arrangement of the particles in the three states.

Student Book pages 80–81 P6

6.3 Changes of state

Key points

- For a pure substance:
  - its melting point is the temperature at which it melts (which is the same temperature at which it solidifies)
  - its boiling point is the temperature at which it boils (which is the same temperature at which it condenses).
- Energy is needed to melt a solid or boil a liquid.
- Boiling occurs throughout a liquid at its boiling point. Evaporation occurs from the surface of a liquid when its temperature is below its boiling point.
- The flat section of a temperature–time graph gives the melting point or the boiling point of a substance.

Synoptic link

For more information on latent heat, look at Topic P6.5.

Key words: melting point, freezing point, boiling point, latent heat

- The temperature at which a solid changes to a liquid is called the **melting point**. This is the same temperature at which the liquid changes to a solid, when it is called the **freezing point**.
- The temperature at which a liquid changes to a gas is called the **boiling point**. This is the same temperature at which a gas changes to a liquid (condenses).
- Boiling takes place throughout a liquid at its boiling point. Evaporation takes place from the surface of a liquid when its temperature is below its boiling point.
- Impurities change the melting point and boiling point of a substance.
- For a substance to change state, energy must be transferred to or from the substance. This energy is called **latent heat**.
- If a graph is plotted of temperature against time for a substance as it is heated, the 'flat' sections of the graph show the melting point and boiling point of the substance. The temperature of the substance does not change whilst the change of state is taking place.
- Water freezes at 0 °C and boils at 100 °C.

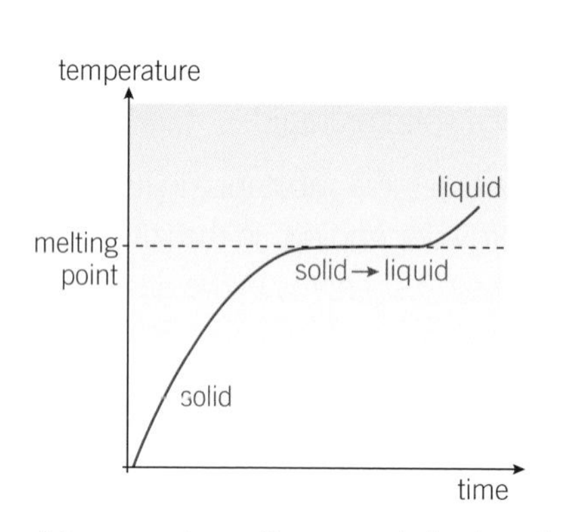

A temperature–time graph for heating

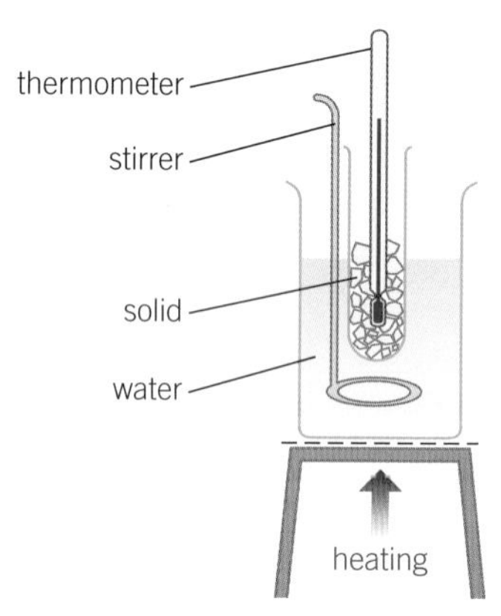

Measuring the melting point of a substance

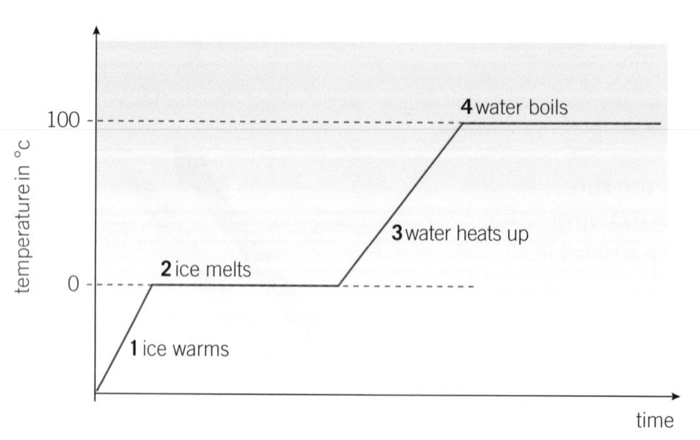

Melting and boiling water

1 What happens to water at its freezing point?
2 What is latent heat?

6.4 Internal energy

Key points

- Increasing the temperature of a substance increases its internal energy.
- The strength of the forces of attraction between the particles of a substance explains why it is a solid, a liquid, or a gas.
- When a substance is heated:
  - if its temperature rises, the kinetic energy of its particles increases
  - if it melts or it boils, the potential energy of its particles increases.
- The pressure of a gas on a surface is caused by the particles of the gas repeatedly hitting the surface.

Synoptic links

For more information on energy transfers, look at Topic P1.1.

For more information on pressure, look at Topic 11.1.

Key word: internal energy

- The **internal energy** of a substance is the total energy in the kinetic energy store and potential energy store of the particles in the substance.
- The particles in a substance have kinetic energy due to their individual motions relative to each other.
- The particles in a substance have potential energy due to their individual positions relative to each other.
- When a substance is heated the energy of its particles increases, so its internal energy increases.
- If the total kinetic energy of the particles increases, the temperature of the substance increases.
- If the substance changes state, the potential energy of its particles increases.
- For a solid:
  - There are strong forces of attraction between the particles and each particle vibrates about a fixed position.
  - When a solid is heated, the particles' energy stores increase and they vibrate more.
- For a liquid:
  - There are weaker forces of attraction between the particles. The forces stop the particles moving completely away from each other, but they are not strong enough to hold the particles in a fixed structure.
  - When a liquid is heated, some particles gain enough energy to break away from the other particles and are in a gas state.
- For a gas:
  - The forces of attraction between the particles are negligible (too small to be noticeable), so the particles of a gas are completely separate from each other.
  - When a gas is heated, its particles gain kinetic energy and move faster.
- The particles in a gas collide with each other and with the walls of their container. During these collisions the particles exert a force, and hence a pressure, on the walls of the container.

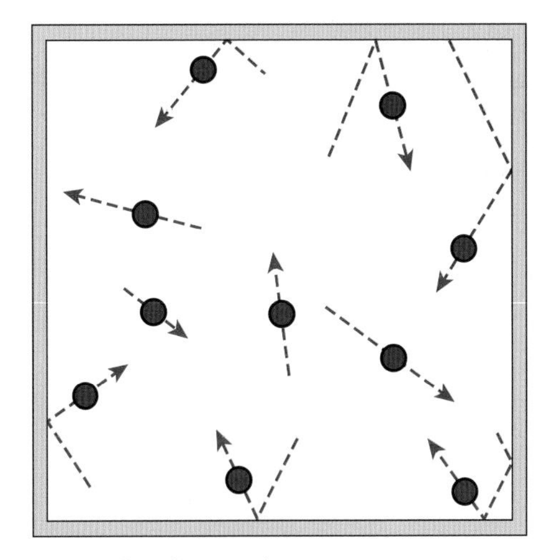

Gas molecules in a box

1. What happens to the internal energy of a substance if it is cooled?
2. In which physical state are the forces of attraction between the particles strongest?

Student Book pages 84–85

P6

6.5 Specific latent heat

Key points

- Latent heat is the energy needed for a substance to change its state without changing its temperature.
- Specific latent heat of fusion (or of vaporisation) is the energy needed to melt (or boil) 1 kg of a substance without changing its temperature.
- In latent heat calculations, use the equation $E = m \times L$
- The specific latent heat of ice (or of water) can be measured using a low-voltage heater to melt the ice (or to boil the water).

Synoptic link

For more information on changes of state, look at Topic P6.3.

Key words: specific latent heat of fusion, specific latent heat of vaporisation

- The latent heat of fusion is the energy required by a substance to change from a solid to a liquid, with no change in temperature. If the substance is cooled, and changes from a liquid to a solid, the latent heat is transferred to the surroundings.
- The **specific latent heat of fusion** is the energy needed to change 1 kg of the substance from a solid to a liquid, with no change in temperature.
- The specific latent heat of fusion can be calculated using the equation $L_F = \frac{E}{m}$, where:

  L_F is the specific latent heat of fusion in joules per kilogram, J/kg

  E is the energy needed in joules, J

  m is the mass in kilograms, kg.
- Latent heat of vaporisation is the energy required by a substance to change from a liquid to a vapour, with no change in temperature. If the substance is cooled, and changes from a vapour to a liquid, the latent heat is transferred to the surroundings.
- The **specific latent heat of vaporisation** is the energy needed to change 1 kg of the substance from a liquid to a vapour, with no change in temperature.
- The specific latent heat of vaporisation can be calculated using the same equation as for specific latent heat of fusion.

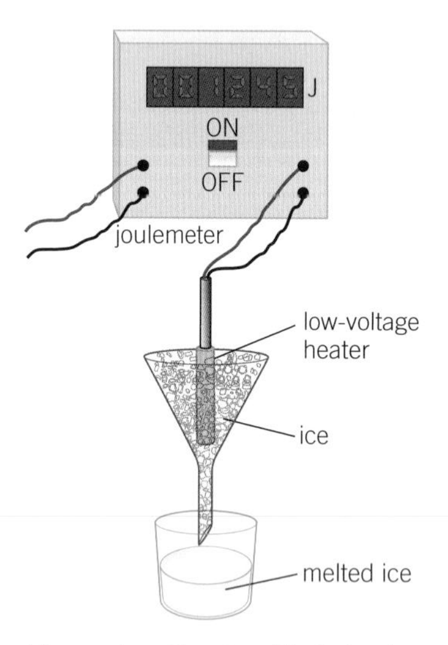

Measuring the specific latent heat of fusion of ice

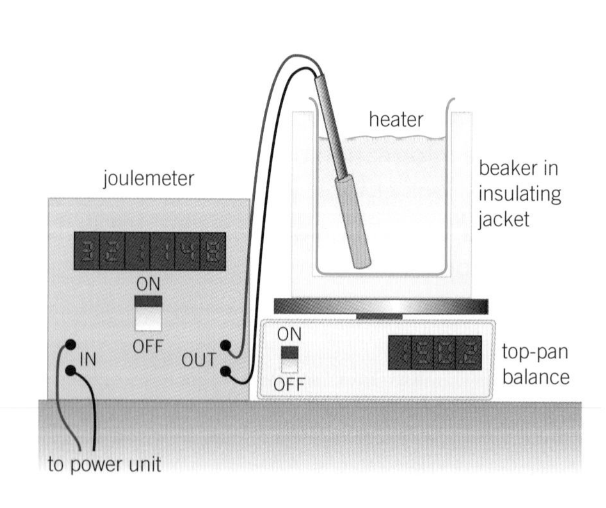

Measuring the specific latent heat of vaporisation of water

- Simple experiments can be used to determine the specific latent heat of fusion and the specific latent heat of vaporisation of water. A low-voltage heater is used to melt ice or boil water. The energy supplied can be measured using a joulemeter.
- Remember that 'specific' in this context indicates that the energy transferred is 'per kilogram'.

1 What is meant by the term 'specific latent heat of vaporisation'?

2 What is the unit of the specific latent heat of fusion?

6.6 Gas pressure and temperature

Key points

- The pressure of a gas is caused by the random impacts of gas molecules on surfaces that are in contact with the gas.
- If the temperature of a gas in a sealed container is increased, the pressure of the gas increases because:
  - the molecules move faster so they hit the surfaces with more force
  - the number of impacts per second of gas molecules on the container increases.
- The unpredictable motion of smoke particles in air is evidence of the random motion of gas molecules.

Synoptic link

For more information on gas pressure, look at Topics P6.4 and P11.1.

On the up

To achieve the top grades, you should be able to explain how the motion of the molecules in a gas is related to its pressure and temperature.

- The particles in a gas move at high speeds in random directions.
- The particles collide with each other and with the walls of their container, rebounding after each collision.
- Each collision exerts a tiny force on the surface of the container and there are millions of collisions every second.
- Pressure is force per unit area, so the total force from all the collisions produces a steady pressure on the walls of the container.
- If the temperature of the gas in a sealed container increases:
  - The particles in the gas move faster and collide with the container walls with more force, causing the pressure to increase.
  - There are more collisions each second with the walls of the container, also causing the pressure to increase.

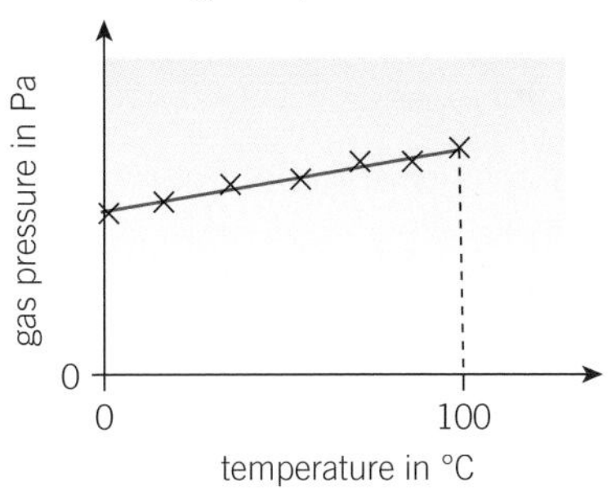

Pressure–temperature graph for a gas

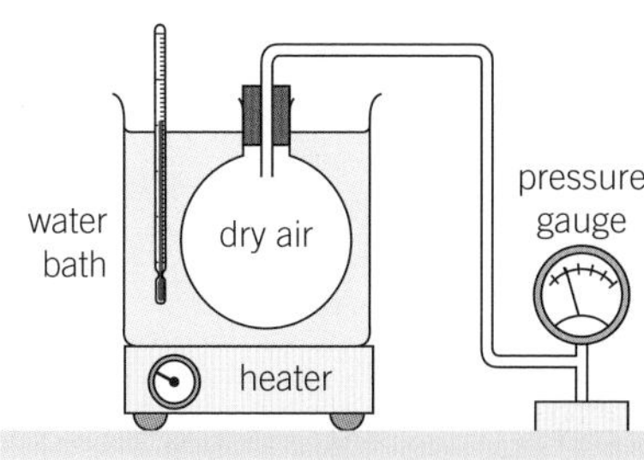

Measuring gas pressure at different temperatures

- You can see how particles in a gas move about randomly by using a microscope to observe smoke particles in a smoke cell. The air particles collide with the smoke particles and cause them to move along unpredictable paths. This random motion is called Brownian motion.

The smoke particle is much larger than the air molecules

The glass cell contains air molecules that are in constant erratic motion. As they collide with the smoke particle they give it a push. The direction of the push changes at random

The random motion of smoke particles

1 What happens to the pressure of a gas in a sealed container if the temperature of the gas decreases?

2 A gas is heated.
What happens to the kinetic energy of the molecules of the gas?

Student Book **pages 88–89** **P6**

~~6.7~~ Gas pressure and volume

Key points

- For a fixed mass of gas at constant temperature:
  - its pressure is increased if its volume is decreased
  - reducing the volume of a gas increases the number of molecular impacts each second on the surfaces that are in contact with the gas.
- Use the equation $pV = \text{constant}$ if the mass and the temperature of the gas do not change.
- (H) The temperature of a gas can increase if it is compressed rapidly because work is done on it and the energy isn't transferred quickly enough to the surroundings.

Synoptic link

For more information on gas pressure, look at Topics P6.4 and P6.6.

- For a fixed mass of gas:
  - The number of gas molecules is constant.
  - If the temperature of the gas is constant, then the average speed of the gas molecules is constant.
- If the volume of a fixed mass of gas is reduced:
  - The molecules have less space to move in so they don't travel so far between collisions.
  - So, the number of collisions per second increases.
  - And so the total force per unit area (that is, the pressure) of the gas increases.

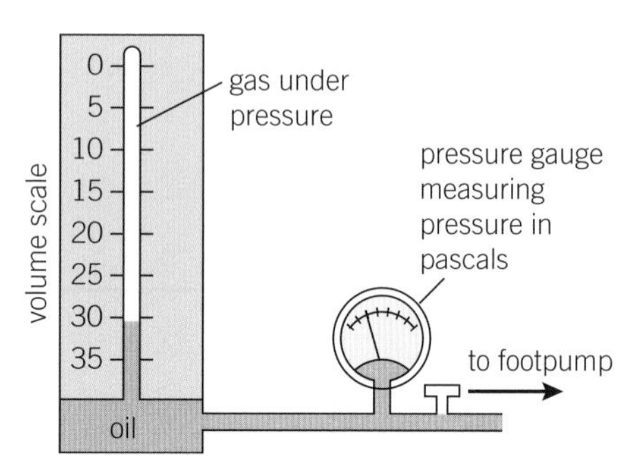

Testing the variation of pressure and volume of a fixed mass of air

- The relationship between pressure and volume is $pV = \text{constant}$, where

  p is pressure in pascals, Pa

  V is volume in cubic metres, m^3.

- This can be rearranged to give

  $$p = \frac{\text{constant}}{V}$$

  This equation shows that the pressure of the gas is inversely proportional to the volume.

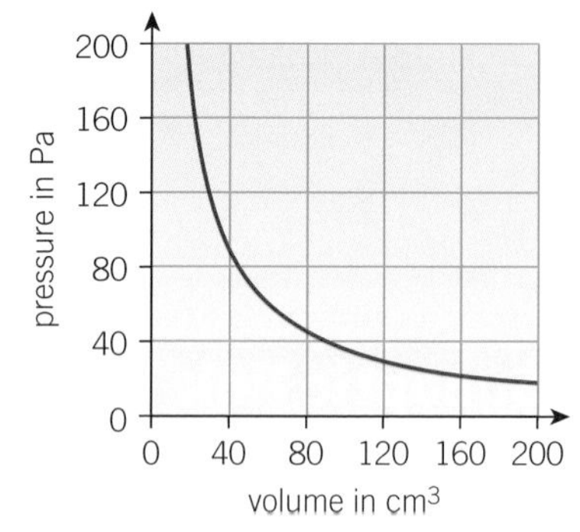

The inversely proportional relationship between volume and pressure

1 What is the unit of pressure?

2 The volume available to a fixed mass of gas at constant temperature increases. What happens to the pressure of the gas?

Higher

If a fixed mass of gas is compressed, work must be done on it. If the gas is compressed slowly, energy is transferred, by heating, to the surroundings. So the internal energy store and the temperature of the gas do not change.

If the gas is compressed quickly, the work done on the gas increases the internal energy store and the temperature of the gas. This is because energy is not transferred quickly enough to the surroundings.

1. Define the density of a substance. [1 mark]
2. Name the most energetic of the states of matter. [1 mark]
3. Explain what is meant by the freezing point and the boiling point of a substance. [2 marks]
4. 0.250 kg of ice was heated until it turned into water. The energy supplied to the ice was 8350 J.
   Calculate the specific latent heat of fusion of water. [2 marks]
5. A steel cube has sides of length 2.00 cm and a mass of 63.2 g.
   Calculate the density of steel in g/cm^3. [2 marks]
6. Describe how the arrangement of particles in a gas is different from that in a solid. [4 marks]
7. Explain, in terms of particles, why a gas exerts a pressure. [4 marks]
8. Sketch a graph to show how the pressure of a fixed mass of gas at constant volume varies with the temperature. [3 marks]
9. A concrete paving slab has a mass of 60 kg and an area of 0.5 m. The density of concrete is 2400 kg/m^3.
   Calculate the thickness of the paving slab. [3 marks]
10. Describe the differences between boiling and evaporation. [4 marks]
11. Describe what is meant by the internal energy of a substance. [3 marks]
12. **H** A sample of gas is stored in a tube with a piston at one end.
    Explain what happens to the pressure and temperature of the gas if the piston is pushed in:
    a slowly
    b quickly. [4 marks]

Chapter checklist

✓

Tick when you have:

reviewed it after your lesson	✓		
revised once – some questions right	✓	✓	
revised twice – all questions right	✓	✓	✓

Move on to another topic when you have all three ticks

6.1 Density	☐	☐	☐
6.2 States of matter	☐	☐	☐
6.3 Changes of state	☐	☐	☐
6.4 Internal energy	☐	☐	☐
6.5 Specific latent heat	☐	☐	☐
6.6 Gas pressure and temperature	☐	☐	☐
6.7 Gas pressure and volume	☐	☐	☐

Student Book
pages 92–93

P7

7.1 Atoms and radiation

Key points

- A radioactive substance contains unstable nuclei that become stable by emitting radiation.
- There are three main types of radiation from radioactive substances – α, β, and γ.
- Radioactive decay is a random event – you can't predict or influence when it will happen.
- Radioactive sources emit α, β, and γ radiation.

- The nuclei of radioactive substances are unstable. They become stable by radioactive decay. In this process they emit radiation.
- The three types of nuclear radiation emitted are **alpha radiation** (α), **beta radiation** (β), and **gamma radiation** (γ).
- It is not possible to predict when an unstable nucleus will decay. It is a random process and is not affected by external conditions. It cannot be predicted, or speeded up by changing external conditions, such as the temperature or pressure.
- Radioactive sources emit α, β, and γ radiation. A particular source may emit more than one type of radiation.

1 Give the names of the three types of nuclear radiation.
2 What happens to the rate of radioactive decay if the temperature doubles?

Key words: alpha radiation, beta radiation, gamma radiation

Student Book
pages 94–95

P7

7.2 The discovery of the nucleus

Key points

- Rutherford used α particles to probe inside atoms. He found that some of the α particles were scattered through large angles.
- The 'plum pudding' model could not explain why some α particles were scattered through large angles.
- An atom has a small, positively charged central nucleus where most of the mass of the atom is located.
- The nuclear model of the atom correctly explained why some α particles scattered through large angles.

1 Why did most alpha particles pass straight through the foil in Rutherford's experiment?
2 Where is most of the mass of an atom concentrated?

- At the start of the 20th century, scientists thought that an atom consisted of a sphere of positive charge with electrons buried inside, like plums in a pudding. This became known as the 'plum pudding' model of the atom.
- Rutherford, Geiger, and Marsden devised an alpha particle scattering experiment, in which they fired alpha particles at very thin gold foil. Alpha particles have a positive charge.
  - Most of the alpha particles passed straight through the foil. This means that most of the atom is empty space.
  - Some of the alpha particles were deflected through small angles. This suggests that the nucleus is charged. Subsequent experiments showed that the charge on the nucleus is positive.
  - A very small number of the alpha particles were deflected through large angles. This suggests that the nucleus has a large mass.

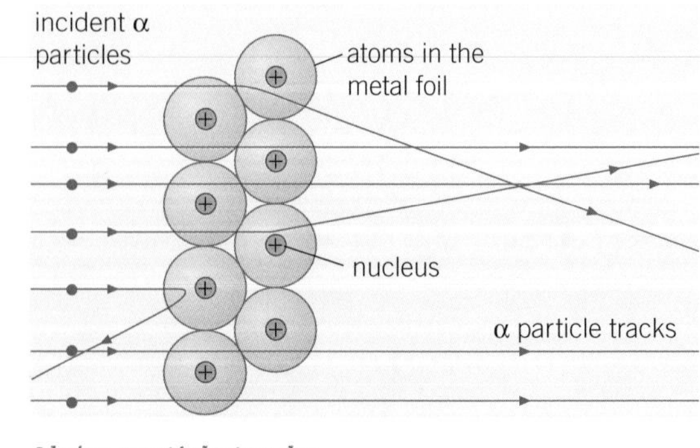

Alpha particle tracks

- The 'plum pudding' model of the atom was not able to explain the results of the alpha particle scattering experiment.
- The 'nuclear model' of the atom was then widely accepted, which says that every atom has a positively charged nucleus, where most of the mass of the atom is concentrated. Electrons orbit the nucleus, and most of the atom is empty space.

Student Book pages 96–97 P7

7.3 Changes in the nucleus

Key points

- Isotopes of an element are atoms with the same number of protons but different numbers of neutrons. So they have the same atomic number as each other but different mass numbers.

	Change in the nucleus	Particle emitted	Equation
α decay	the nucleus loses 2 protons and 2 neutrons	2 protons and 2 neutrons emitted as an α particle	${}^{A}_{Z}X = {}^{A-4}_{Z-2}Y + {}^{4}_{2}\alpha$
β decay	a neutron in the nucleus changes into a proton and an electron	the electron is instantly emitted from the nucleus	${}^{A}_{Z}X = {}^{A}_{Z+1}Y + {}^{0}_{-1}\beta$

Key words: atomic number, mass number, isotopes

Study tip

To help you learn this topic, practise writing nuclear equations for alpha and beta decay.

- The number of protons in an atom is its **atomic number**. It has the symbol Z.
- The total number of protons and neutrons in an atom is its **mass number**. It has the symbol A.
- All atoms of a particular element have the same number of protons. Atoms of the same element with different numbers of neutrons are called **isotopes**.

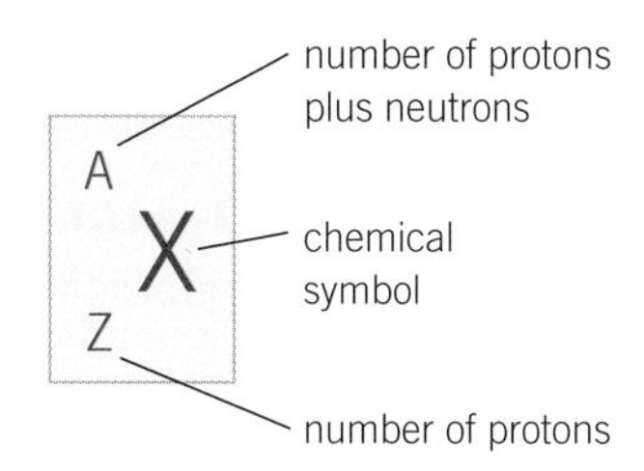

example: the symbol for the uranium isotope with 92 protons and 146 neutrons is

${}^{238}_{92}U$ (or sometimes U-238)

Representing an isotope

- An alpha particle consists of two protons and two neutrons. Its relative mass is 4 and its relative charge is +2. We represent it as ${}^{4}_{2}\alpha$.
- When an unstable nucleus emits an alpha particle the atomic number goes down by 2, and the mass number goes down by 4.
- A beta particle is a high-speed electron from the nucleus, emitted when a neutron in the nucleus changes to a proton and an electron. Its relative mass is 0 and its relative charge is –1. We represent it as ${}^{0}_{-1}\beta$.
- The proton stays in the nucleus, so the atomic number goes up by 1. The mass number is unchanged. The electron is instantly emitted.
- When a nucleus emits gamma radiation there is no change in the atomic number or the mass number. A gamma ray is an electromagnetic wave released from the nucleus. It has no charge and no mass.
- Neutrons are emitted by some radioactive substances. Neutrons are uncharged.

1 What is the relative charge of an alpha particle?

2 What is the relative charge of a beta particle?

Student Book
pages 98–99

P7

7.4 More about alpha, beta, and gamma radiation

Key points

- α radiation is stopped by paper and has a range of a few centimetres in air. It consists of particles, each composed of two protons and two neutrons. It has the greatest ionising power.
- β radiation is stopped by thin sheet of metal and has a range of about one metre in air. It consists of fast-moving electrons emitted from the nucleus. It is less ionising than alpha radiation, and more ionising than gamma radiation.
- γ radiation is stopped by thick lead and has an unlimited range in air. It consists of electromagnetic radiation.
- Alpha, beta, and gamma radiation ionise substances they pass through. Ionisation in a living cell can damage or kill the cell.

Key words: ionisation, irradiated, radioactive contamination

Study tip

To help you remember the properties of alpha, beta, and gamma radiation, practise drawing labelled diagrams to represent their structure, relative ionising power, and penetrating range.

- The radiation from radioactive substances can knock electrons out of atoms. The atoms become positively charged because they lose electrons. This is called **ionisation**.
- Ionisation in a living cell can damage or kill the cell.
- An object becomes **irradiated** if it is exposed to ionising radiation. Remember that the object itself does not become radioactive.
- **Radioactive contamination** is the unwanted presence of materials containing radioactive atoms on other materials.
- Alpha particles consist of two protons and two neutrons. Because they are relatively large, alpha particles have lots of collisions with atoms – they are strongly ionising.
- Because of these collisions, the alpha particles do not penetrate far into a material. They can be stopped by a thin sheet of paper, human skin, or a few centimetres of air.
- Beta particles are fast-moving electrons emitted from the nucleus. As they are much smaller and faster than alpha particles, they are less ionising and penetrate further into a material than alpha particles.
- Beta particles are blocked by a thin sheet of aluminium or a few metres of air.
- Gamma rays are electromagnetic waves so they will travel a long way through a material before colliding with an atom. They are weakly ionising and very penetrating.
- Several centimetres of lead or several metres of concrete are needed to absorb most of the radiation from a gamma source. Gamma radiation has an unlimited range in air.

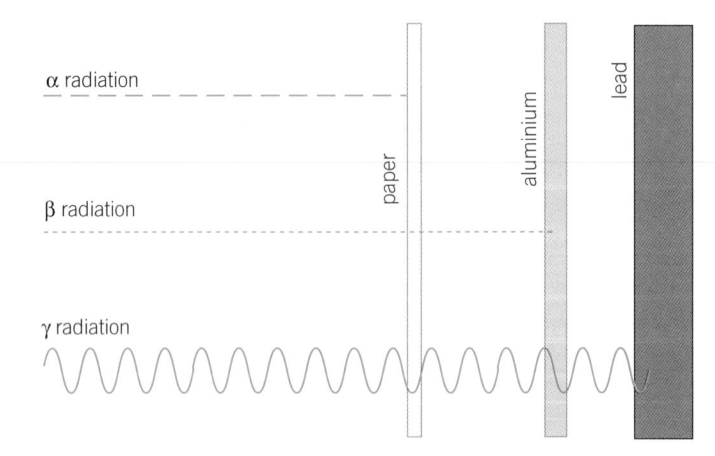

The penetrating power of alpha, beta, and gamma radiation

1 Which type of nuclear radiation is the least penetrating?

2 Which type of nuclear radiation if the least ionising?

7.5 Activity and half-life

Key points

- The half-life of a radioactive isotope is the average time it takes for the number of nuclei of the isotope in a sample to halve.
- The count rate of a Geiger counter caused by a radioactive source decreases as the activity of the source decreases.
- The number of atoms of a radioactive isotope and the count rate both decrease by half every half-life.
- (H) The count rate after n half-lives = the initial count rate ÷ 2^n.

Synoptic link

For more information on isotopes and radioactive decay, look at Topics P7.1 and P7.3.

Key words: activity, count rate, half-life

Study tip

Make sure that you practise doing half-life calculations.

On the up

To achieve the top grades, you need to be able to determine the half-life of a radioisotope from a graph and carry out half-life calculations.

- The **activity** of a radioactive source is the number of unstable atoms in the source that decay per second. The unit of activity is the Becquerel (Bq), which is 1 decay per second.
- We can measure the radioactivity of a sample of a radioactive material by measuring the **count rate** from it. The count rate is the number of counts per second.
- The radioactivity of a sample decreases over time. How quickly the count rate falls to nearly zero depends on the material. Some take a few minutes, others take millions of years. We use the idea of **half-life** to measure how quickly the radioactivity decreases.
- The half-life of a radioactive isotope is the average time it takes:
  - for the number of nuclei of the isotope in a sample (and so the mass of parent atoms) to halve
  - for the count rate from the isotope to fall to half its initial value.

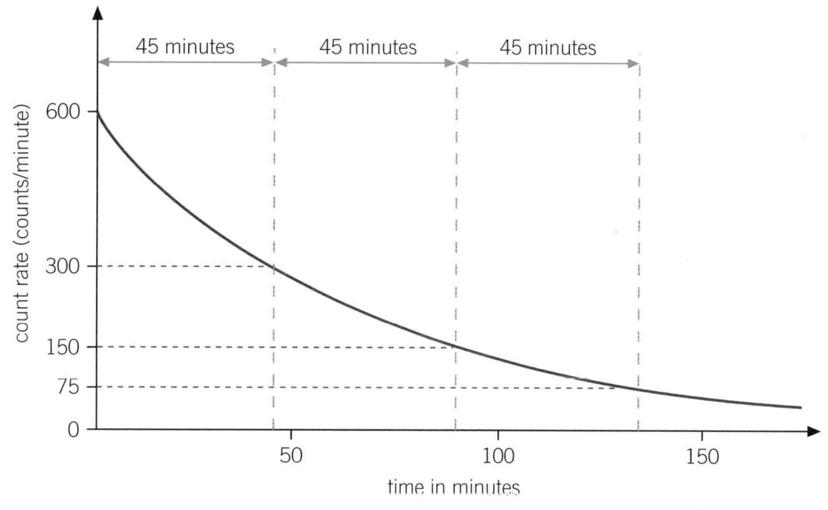

A graph of count rate against time

- Radioactive decay is a random process, so it is not possible to predict when a particular nucleus will decay.

1. What happens to the count rate of a radioactive sample over time?
2. What has happened to the original count rate of a radioactive sample after two half-lives have passed?

Higher

Half-life calculations

$$\text{count rate after } n \text{ half-lives} = \frac{\text{initial count rate}}{2^n}$$

or

$$\text{number of unstable nuclei after } n \text{ half-lives} = \frac{\text{initial number of unstable nuclei}}{2^n}$$

Student Book **pages 102–103**

P7 7.6 Nuclear radiation in medicine

Key points

- Radioactive isotopes are used in medicine for medical imaging and treatment of cancer, and as tracers to monitor organs.
- How useful a radioactive isotope is depends on:
  - its half-life
  - the type of radiation it gives out.
- For medical imaging with a radioactive isotope and for medical tracers, the half-life should be not too short and not too long.
- A gamma beam or a radioactive implant can destroy cancer cells in a tumour.

Synoptic links

For more information on using gamma radiation in cancer treatment, look at Topic P13.4.

For more information on medical imaging, look at Topics P12.6 and P13.5.

Study tip

In a table, relate the properties of beta and gamma radiation to their uses in medicine.

On the up

To achieve the top grades, you should be able to evaluate the use of different nuclear radiations for use in diagnosis and treatment.

- Radioactive tracers can be used to measure the flow of a substance through an organ. The patient is injected with (or drinks) a radioactive isotope that emits gamma radiation. The movement of the radioactive isotope through the organ is monitored using a gamma detector placed outside the body.
- Gamma cameras take pictures of internal organs. The patient is injected with a radioactive isotope that emits gamma radiation. The isotope is absorbed by the organ and a gamma camera outside the body, connected to a computer, detects the radiation and uses it to build up an image of the organ.
- In both cases, the half-life of the radioactive isotope used must be long enough to give a useful image of the organ, but must decay (almost completely) shortly after the image has been taken.

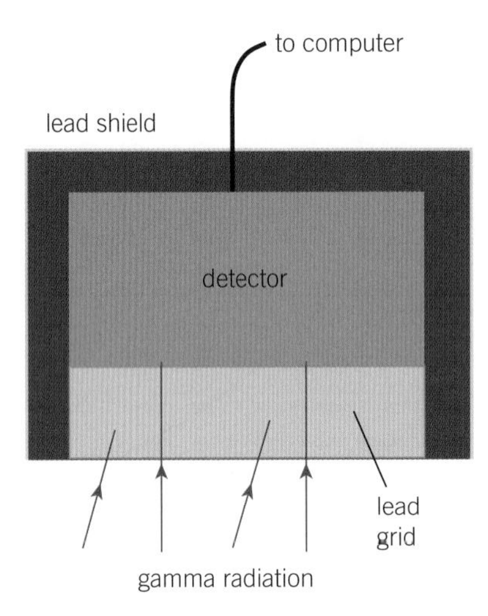

The gamma camera

- A narrow beam of gamma radiation can be used to destroy cancerous tumours. Gamma radiation is used because it can penetrate deep into the body.
- Radioactive implants in the form of small seeds or rods are placed into cancerous tumours.
  - The implants contain beta- or gamma-emitting isotopes.
  - The isotope used must have a half-life long enough to complete the imaging or treatment, but short enough to avoid exposing the patient to unnecessary radiation.

1. Why don't hospitals use alpha sources as a tracer?
2. Why do medical tracers have half-lives of just a few hours?

7.1 Nuclear fission

Key points

- Nuclear fission is the splitting of an atom's nucleus into two smaller nuclei and the release of two or three neutrons and energy.
- Induced fission occurs when a neutron is absorbed by a uranum-235 nucleus or a plutonium-239 nucleus and the nucleus splits. Spontaneous fission occurs without a neutron being absorbed.
- A chain reaction occurs in a nuclear reactor when each fission event causes further fission events.
- In a nuclear reactor, control rods absorb fission neutrons to ensure that, on average, only one neutron per fission goes on to produce further fission.

Synoptic link

For more information on nuclear reactors, look at Topic P7.9.

Study tip

Practise drawing a simple diagram to represent a chain reaction.

Key words: nuclear fission, chain reaction, nuclear reactor, reactor core, moderator

1 What is enriched uranium?
2 What is induced fission?

- **Nuclear fission** is the splitting of an atomic nucleus into two smaller nuclei.
- The fission process results in the emission of:
  - two or three high-speed neutrons
  - energy – in the form of gamma radiation, and kinetic energy of the nuclei and neutrons.
- Naturally occurring uranium is mostly uranium-238, which is non-fissionable. Most nuclear reactors use 'enriched' uranium that contains 2–3% uranium-235, which is fissionable. Some reactors use plutonium-239.
- For fission to occur, the uranium-235 or plutonium-239 nucleus must absorb a neutron. This is called induced fission. Very rarely, fission occurs without a neutron being absorbed. This is called spontaneous fission.
- A **chain reaction** occurs when the neutrons released from each fission event cause further fission events to occur.

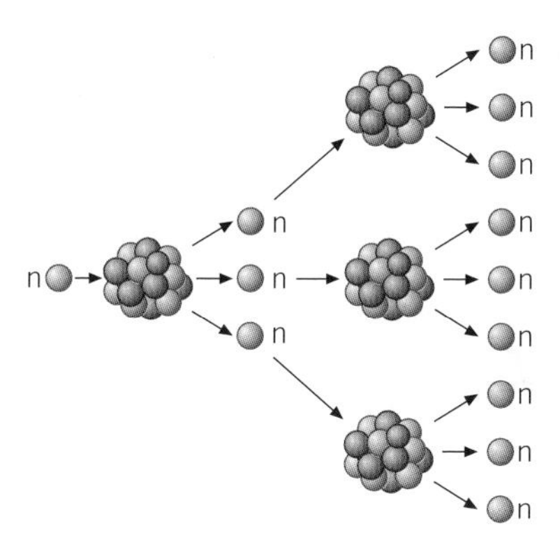

A chain reaction in a nuclear reactor

- In a **nuclear reactor** the fission process is controlled, so one fission neutron per fission, on average, goes on to produce further fission. Control rods in the **reactor core** absorb surplus neutrons. The depth of the rods in the core is adjusted to maintain a steady chain reaction.
- Fast neutrons don't cause further fission, so in a reactor core a **moderator** is used to slow the high-speed neutrons.

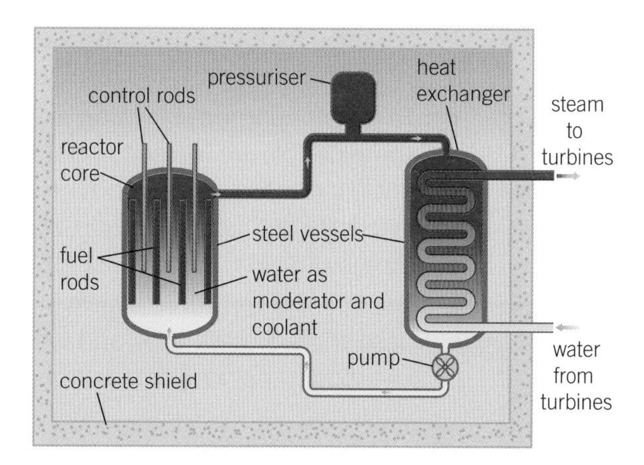

A nuclear reactor

Student Book
pages 106–107

P7

~~7~~.8 Nuclear fusion

Key points

- Nuclear fusion is the process of forcing the nuclei of two atoms close enough together so that they form a single larger nucleus.
- Nuclear fusion can be brought about by making two light nuclei collide at very high speed.
- Energy is released when two light nuclei are fused together. Nuclear fusion in the Sun's core releases energy.
- A fusion reactor needs to be at a very high temperature before nuclear fusion can take place. The nuclei to be fused are difficult to contain.

- **Nuclear fusion** is the process of forcing two nuclei close enough together so that they form a single larger nucleus.
- During this process, some of the mass of the small nuclei is converted to energy.
- Nuclear fusion can be brought about by making two light nuclei collide at very high speed.
- Nuclear fusion is the process by which energy is released in stars, including the Sun.
- There are very big technical difficulties with making fusion a useful source of energy. Nuclei approaching each other will repel one another due to their positive charge. To overcome this, the nuclei must be heated to very high temperatures to give them enough kinetic energy to overcome the repulsion, and fuse together. Because of the enormously high temperatures involved, the reaction cannot take place in a normal 'container', since the nuclei would touch the edges, grow cold, and fusion would stop. Instead, the fusion reaction has to be contained by a magnetic field.

1 By what process is energy released in the Sun?
2 How are nuclei contained in a fusion reactor?

Key word: nuclear fusion

Student Book
pages 108–109

P7

~~7~~.9 Nuclear issues

Key points

- Radon gas is an α-emitting isotope that seeps into houses through the ground in some areas.
- There are hundreds of fission reactors safely in use around the world. None of them is of the same type as the Chernobyl reactors that exploded.
- Nuclear waste contains many different radioactive isotopes that emit nuclear radiation for many years. The radiation is dangerous because it can cause cancer.
- Nuclear waste is stored in safe and secure conditions for many years after unused uranium and plutonium (to be used in the future) are removed from it.

- High levels of background radiation can be harmful as it ionises substances it passes through. However, nuclear power contributes very little to background radiation.
- The major source of background radiation in the air is radon gas, which seeps through the ground from radioactive substances in rocks deep underground. Radon gas emits alpha particles, so it is a health hazard if breathed in. Other sources of background radiation include cosmic rays from outer space, food and drink, air travel, nuclear weapons testing, and medical applications such as X-rays.
- Used fuel rods from nuclear power stations contain lots of radioactive isotopes with long half-lives. After any unused uranium and plutonium has been removed for future use, the rods must be stored in secure conditions for many years until their radiation levels are safe.
- There are hundreds of fission reactors safely in use throughout the world, but some people are worried about accidents, such as the one that occurred at the nuclear reactors in Chernobyl in 1986. However, none of the reactors currently in use are of this type. Lessons learned from Chernobyl have contributed to the design of safer nuclear power stations.
- The effect of radiation on living cells depends on the type of radiation, the dose, the exposure time, and whether the source is inside or outside the body. The bigger the dose of radiation, the higher the risk of cancer.

1 What are the two major sources of background radiation for a person living in the UK?
2 Why must radioactive waste be stored securely?

1. Describe the effect of pressure on the rate of radioactive decay. [1 mark]
2. Name what most of an atom consists of. [1 mark]
3. Define what is meant by the atomic number and the mass number of an atom. [2 marks]
4. Describe what nuclear fusion is. [2 marks]
5. Determine how many protons and how many neutrons there are in the nucleus of the isotope $^{239}_{94}Pu$. [2 marks]
6. Describe the differences between alpha, beta, and gamma radiation in terms of their penetrating abilities. [6 marks]
7. A radioactive isotope has a half-life of 8 days. A sealed sample contains 24 milligrams of the isotope.
   Calculate the mass of the isotope in the sample after 32 days. [3 marks]
8. Draw a diagram to show what is meant by a chain reaction. [3 marks]
9. A sample of a radioactive isotope contains 800 000 radioactive nuclei. The isotope has a half-life of 15 hours.
   Calculate the number of radioactive nuclei remaining after 45 hours. [3 marks]
10. A particular isotope of cobalt is a gamma emitter with a half-life of 5 years.
    Explain why this isotope is suitable to use for treating cancer. [3 marks]
11. Explain the difference between induced fission and spontaneous fission. [3 marks]
12. Explain the function of the moderator and the control rods in a nuclear reactor. [4 marks]

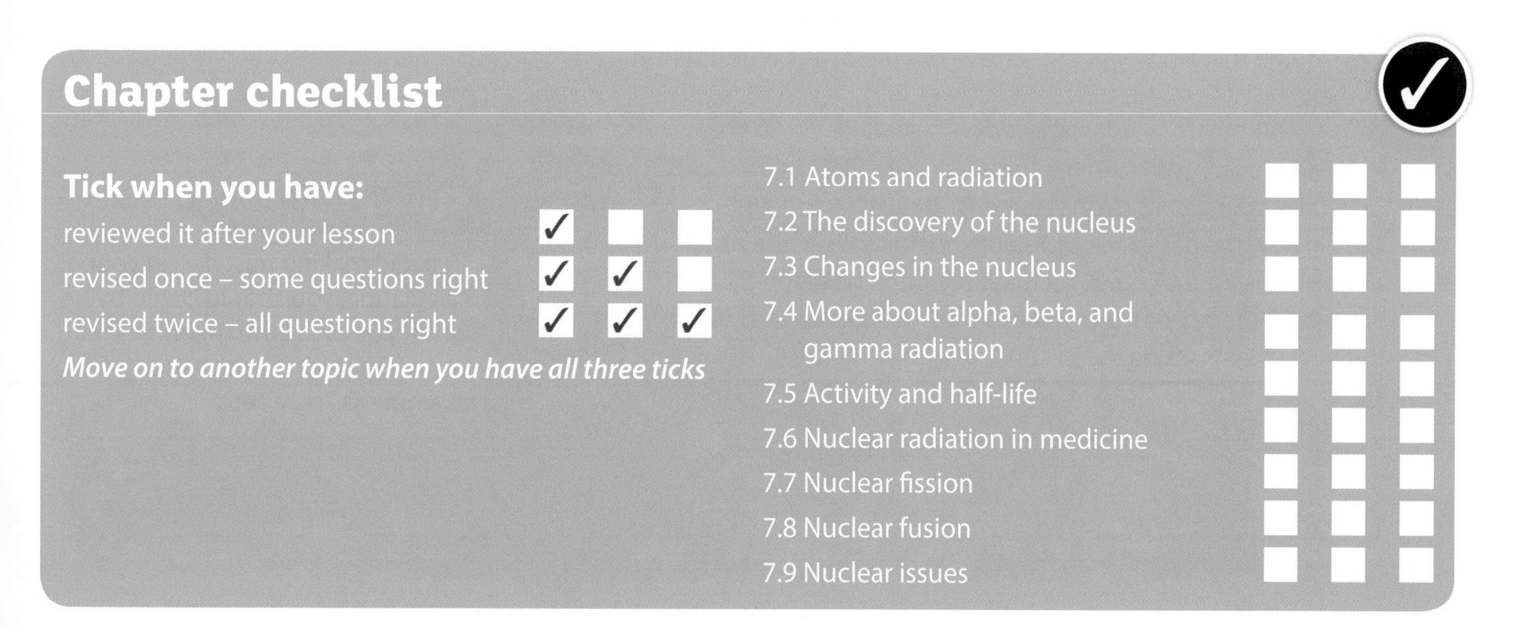

01 A student sets up the circuit shown in **Figure 1**.

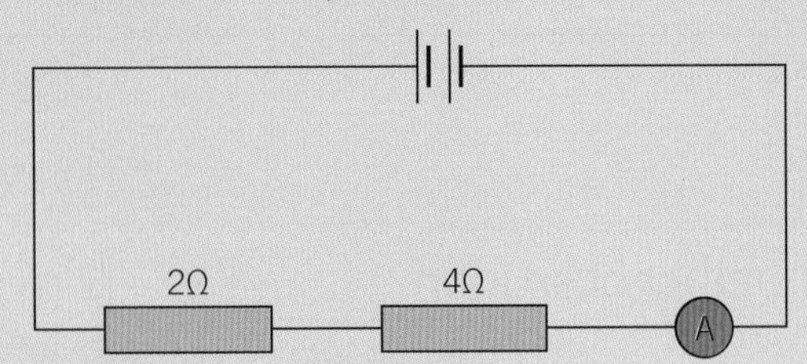

Figure 1

The reading on the ammeter is 0.5 A.

01.1 Calculate the potential difference across the 2 Ω resistor. [3 marks]

01.2 The student measures the potential difference across the 4 Ω resistor using a voltmeter. Copy the diagram and add the correct circuit symbol to show how the student should connect the voltmeter. [2 marks]

01.3 The potential difference across the 4 Ω resistor is 2 V. Calculate the potential difference across the battery. [1 mark]

02 Alpha, beta, and gamma are types of nuclear radiation.

02.1 Which type of nuclear radiation is stopped by a sheet of paper? [1 mark]

02.2 Which type of nuclear radiation is an electromagnetic wave? [1 mark]

02.3 Which type of nuclear radiation has a positive charge? [1 mark]

02.4 The isotope uranium-238 emits alpha radiation. The isotope can be represented as $^{238}_{92}U$. How many protons and how many neutrons are there in an atom of $^{238}_{92}U$? [2 marks]

03 A student is provided with a beaker containing a liquid. Describe the method she should use to determine the density of the liquid. [5 marks]

04 In an investigation, a low-voltage heater was used to melt some ice.

04.1 The potential difference across the heater was 50 V and the current through it was 4.2 A. Calculate the power of the heater. Give the unit. [4 marks]

04.2 The energy supplied by the heater was 6300 J. The mass of the ice melted was 100 g. Calculate the specific latent heat of fusion of the ice. Assume that all the energy supplied by the heater is transferred to the ice. [3 marks].

05 A sealed piston containing gas has a volume of 1200 cm^3. The pressure of the gas inside the piston is 101 kPa.

05.1 Explain how the gas exerts a pressure on the walls of the piston. [3 marks]

05.2 The piston is used to slowly increase the pressure of the gas to 303 kPa. The temperature of the gas does not change. Calculate the new volume inside the piston. [2 marks]

05.3 (H) Explain why the temperature of the gas only remains constant if the pressure is increased slowly. [3 marks]

Study tip

To avoid making mistakes in your calculation, you should begin by writing down all the quantities that you are given in the question – for example, p_1, p_2, V_1, etc.

06 Medical instruments may be sterilised by irradiating them using a source of gamma radiation. This process kills any microorganisms on the instruments. The instruments are sealed in airtight plastic bags, placed on a conveyor belt, and then passed through a chamber containing the gamma source (as shown in **Figure 2**).

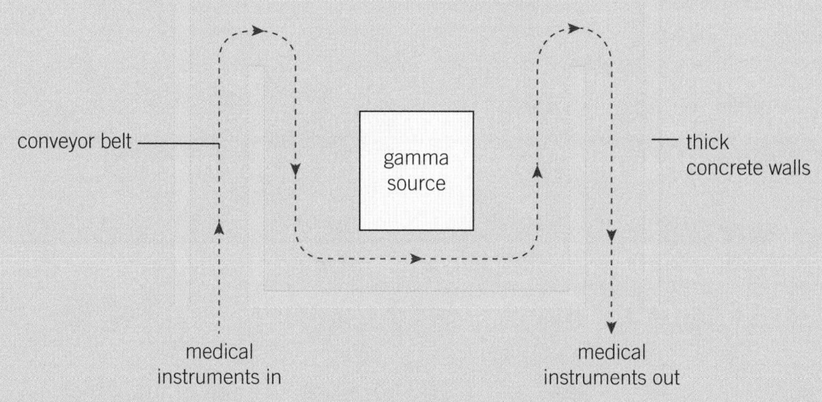

Figure 2

The conveyor belt stops at 120 positions in the chamber. The time it stops at each position depends on the density of the instrument. The table shows the relative densities of different instruments, the time the conveyor stops at each position, and the total irradiation time for an instrument of that density.

Relative density of instrument	Time stopped at each position in seconds	Total irradiation time in hours
1	208	6.93
2	230	7.67
3	256	8.53
4	283	

06.1 Complete the table by calculating the total irradiation time for a product with relative density 4. [3 marks]

06.2 Explain the relationship between relative density and total irradiation time. [3 marks]

06.3 Explain why the instruments are sealed in airtight plastic bags. [2 marks]

06.4 Explain why the chamber containing the gamma source has walls made from thick concrete. [2 marks]

06.5 In some countries, gamma radiation is used to sterilise food. In the UK there is much opposition to this and food that has been irradiated must be labelled. Evaluate the use of gamma radiation to sterilise food. [6 marks]

Study tip

In an 'evaluate' question you should use the information given in the question, as well as your own knowledge, to consider evidence for and against the issue in question.

3 Forces in action

An astronaut in a space station can float around and perform acrobatic tricks. Many people think this is because there is no gravity in space. However, this is wrong. The force of gravity due to the Earth stretches far into space and keeps the space station orbiting the Earth.

You and all of the objects around you are acted on by the force of gravity. You are also acted on by other forces, such as friction, which acts between objects when they touch each other, and non-contact forces like magnetic and electrostatic forces.

In this section, you will learn about what forces do, how we measure them and their effects, and how we calculate the effect forces have on objects.

I already know...	I will revise...
force is measured in newtons (N) using a newton-meter.	the difference between a vector and a scalar and how to represent a vector.
an object is in equilibrium when the forces acting on it are balanced.	how to find the resultant of two forces and to resolve a force into perpendicular components.
speed is measured in metres per second.	the difference between speed and velocity and what we mean by acceleration.
drag forces and friction resist the motion of moving objects.	what is meant by terminal velocity and why objects fall through water at a constant velocity.
when objects interact, each one exerts a force on the other.	what is meant by conservation of momentum and when we can use this rule.
the force in a stretched object is called tension and it increases if the object is stretched more.	how to measure the stiffness of a spring and what is meant by elasticity.
the weight of an object is due to the force of gravity on it.	how to calculate the weight of an object from its mass and the gravitational field strength of where it is.

Student Book **pages 114–115** **P8**

8.1 Vectors and scalars

Key points

- Displacement is distance in a given direction.
- A vector quantity is a physical quantity that has magnitude and direction.
- A scalar quantity has magnitude but no direction.
- A vector quantity can be represented by an arrow in the direction of the vector and of length in proportion to the magnitude of the vector.

- **Scalar** quantities have **magnitude** (size) but no direction. Examples include distance, speed, time, mass, energy, and power.
- **Vector** quantities have both magnitude and direction. Examples include displacement, velocity, force, weight, and momentum.
- **Displacement** is distance travelled in a given direction.
- Vectors may be represented by a diagram. An arrow is drawn in the direction of the vector and with a length in proportion to the magnitude of the vector.

1 What is the difference between a scalar and a vector?

2 An arrow is drawn with a scale of 2 cm to represent 5 N. What length of arrow should be drawn to represent a 17.5 N force?

Key words: scalar, magnitude, vector, displacement

Student Book **pages 116–117** **P8**

8.2 Forces between objects

Key points

- A force can change the shape of an object or change its motion or state of rest.
- The unit of force is the newton (N).
- A contact force is a force that acts on objects only when the objects touch each other.
- When two objects interact they always exert equal and opposite forces on each other.

Synoptic link

For more information on Newton's laws, look at Topic P10.1.

Key words: force, Newton's third law of motion, driving force, friction

Study tip

Draw diagrams showing examples of Newton's third law of motion in action to help you remember them.

- A **force** is a push or pull that acts on an object due to its interaction with another object. Forces are measured in newtons, abbreviated to N.
- A force can change the shape of an object, start or stop it moving, or change its direction.
- If two objects must touch each other to interact, the forces between them are called contact forces. Examples of contact forces are friction and air resistance.
- **Newton's third law of motion** states that when two objects interact with each other, they exert equal and opposite forces on each other. These force pairs are sometimes called 'action and reaction' forces. The two forces are always of the same type (e.g., both frictional forces), and each force always acts on a different object from its partner. For example:
  - If a car hits a barrier, it exerts a force on the barrier. The barrier exerts a force on the car that is equal in size and in the opposite direction.
  - If you place a book on a table, the contact force of the book will act vertically downwards on the table. The table will exert an equal and opposite reaction force upwards on the book.
  - When a car is being driven forwards there is a force from the tyre on the ground pushing backwards. This is called the **driving force**. There is an equal and opposite force from the ground on the tyre that pushes the car forwards. The force between the tyre and the ground is caused by **friction**. Friction acts where there is contact between surfaces.

An example of Newton's third law of motion

1 What is the SI unit of force?

2 In what direction does the force of weight always act?

Student Book pages 118–119

P8

8.3 Resultant forces

Key points

- The resultant force is a single force that has the same effect as all the forces acting on an object.
- If the resultant force on an object is:
  - zero, the object stays at rest or at the same speed and direction
  - greater than zero, the speed or direction of the object will change.
- If two forces act on an object along the same line, the resultant force is:
  - their sum, if the forces act in the same direction
  - their difference, if the forces act in opposite directions.
- (H) A free-body force diagram of an object shows the forces acting on it.

Study tip

Practise drawing free-body force diagrams for objects in different situations.

Key words: resultant force, Newton's first law of motion

- Most objects have more than one force acting on them. The **resultant force** on an object is the single force that has the same effect as all the forces acting on the object.
- **Newton's first law of motion** states that if the forces acting on an object are balanced, the resultant force on the object is zero and:
  - if the object is at rest, it will stay at rest
  - if the object is moving, it will carry on moving at the same speed and in the same direction.
- When the resultant force on an object is not zero, the forces acting on the object are not balanced. The movement of the object depends on the size and direction of the resultant force.
- If two forces act on an object along the same line of motion, the resultant force will be:
  - the difference between the forces (if they act in opposite directions) and in the direction of the larger force
  - the sum of the forces (if they act in the same direction) and in the same direction as both forces.

1 What is the magnitude and direction of the resultant force when a 7 N force and a 3 N force act in the same direction on an object?

2 What is the magnitude and direction of the resultant force when a 7 N force and a 3 N force act in opposite directions on an object?

Force diagrams

Higher

A free-body force diagram show all of the forces acting on a single object. The forces are represented by arrows. The direction of the arrow shows the direction of the force and the length of the arrow, drawn to scale, represents the magnitude (size) of the force.

Student Book pages 120–121 P8

8.4 Moments at work

Key points

- The moment of a force is a measure of the turning effect of the force on an object.
- The moment of a force about a pivot is $M = F \times d$, where d is the perpendicular distance from the line of action of the force F to the pivot.
- To increase the moment of a force, increase F or increase d.
- Levers can be used to exert a force that is greater than the effort.

Key words: moment, load

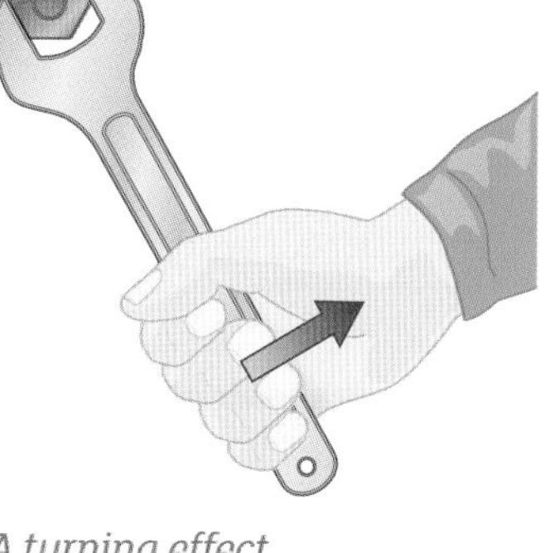

A turning effect

Study tip

Look out for everyday examples of using moments – such as opening a door, using a screwdriver to open a can of paint, or pushing down on the handle of a buggy to lift the front wheels up a kerb. For each example, identify where the force is applied and the position of the pivot.

- The turning effect of a force is called its **moment**.
- The size of the moment is given by the equation $M = F \times d$, where:

  M is the moment of the force in newton metres, N m

  F is the force in newtons, N

  d is the perpendicular distance from the line of action of the force to the pivot in metres, m.

- You can increase the size of the moment by:
  - increasing the magnitude of the force
  - increasing the perpendicular distance from the line of the force to the pivot.

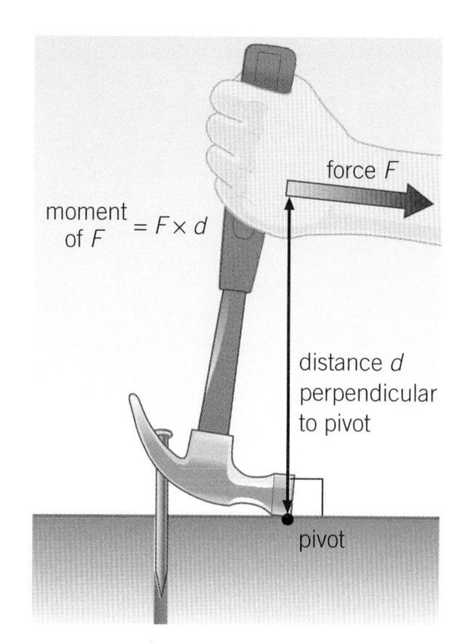

Using a claw hammer. The claw hammer is being used to remove a nail from a wooden beam

- It is easier to undo a wheel nut by pushing on the end of a long spanner than a short one. That's because the long spanner increases the perpendicular distance between the line of action of the force and the pivot.
- You use a lever to move a heavy object. The object is called the **load** and the force you apply to the lever is called the effort. The effort needed to lift the load is only a fraction of the object's weight. A lever acts as a force multiplier – the effort moves a much bigger load.

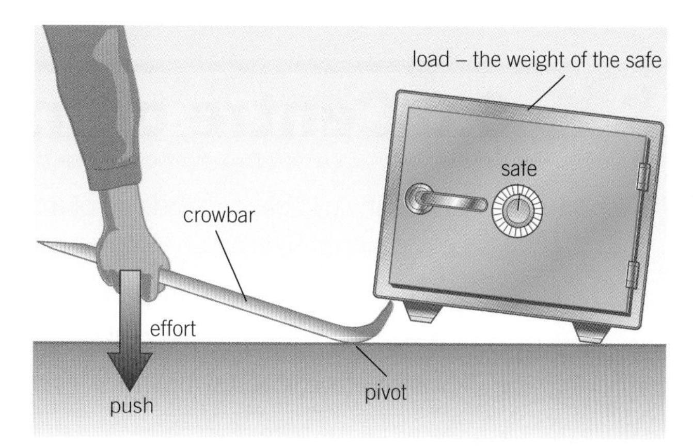

Using a crowbar

1. A door opens when you apply a force of 20 N at right angles to it, 0.6 m from the hinge. What is the moment of the force about the hinge?
2. What force would be needed to open the door if it were applied at right angles and 0.3 m from the hinge?
3. How could the moment applied by the claw hammer to the nail in the diagram above be increased?

Student Book **pages 122–123** **P8**

8.5 More about levers and gears

Key points

- A lever used as a force multiplier exerts a greater force than the force applied to the lever by the effort.
- The pivot of a force multiplier is nearer to the line of action of the force it exerts than to the force applied to it.
- Gears are used to change the moment of a turning effect.
- To increase the moment of a turning effect, a small gear wheel needs to drive a large gear wheel.

- A lever can be used to increase the size of a force acting on an object or to make the object turn more easily.

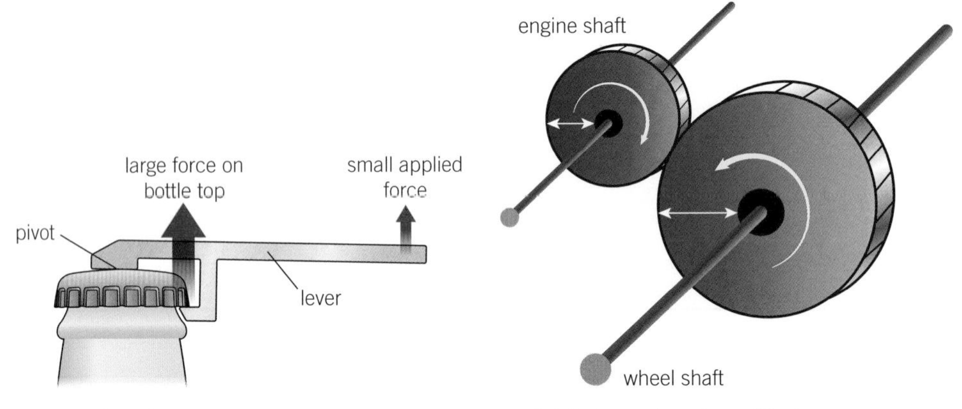

Using a bottle opener

Multiplying a turning effect

- When you use a lever, the force you apply to the lever is further away from the pivot, and smaller in size, than the force that the lever applies on the object. The lever is then called a force multiplier.
- Gears are like levers because they can multiply the effect of a turning force.
  - A low gear gives low speed and a high turning effect.
  - A high gear gives high speed and a low turning effect.

1 What are gears used for?
2 What happens if a large gear wheel is used to drive a small gear wheel?

Student Book **pages 124–125** **P8**

8.6 Centre of mass

Key points

- The centre of mass of an object is the point where its mass can be thought of as being concentrated.
- The centre of mass of a uniform ruler is at its midpoint.
- When an object is freely suspended, it comes to rest with its centre of mass directly underneath the point of suspension.
- The centre of mass of a symmetrical object is along the axis of symmetry.

- Although any object is made up of many particles, its mass can be thought of as being concentrated at one single point. This point is called the centre of mass.
- Any object that is freely suspended will come to rest with its centre of mass directly below the point of suspension. The object is then in equilibrium.
- For a flat object that is symmetrical, its centre of mass lies along the axis of symmetry.
- The wider the base of an object and the lower its centre of mass, the more stable it is.

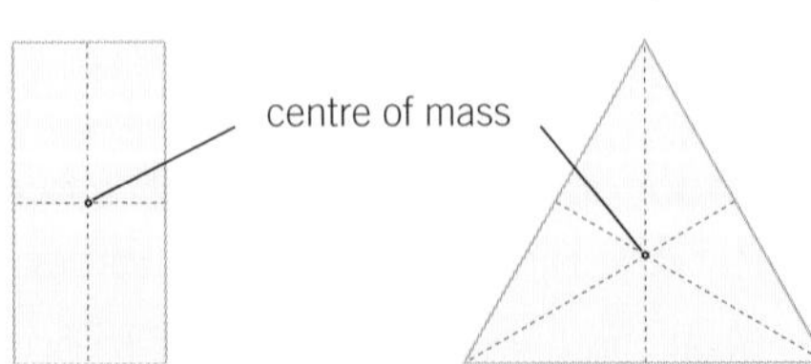

Symmetrical objects

1 Where is the centre of mass of a symmetrical object?
2 Why do racing cars have a wide base and a low centre of mass?

8.7 Moments and equilibrium

Key points

- If an object at rest doesn't turn, the sum of the anticlockwise moments about any point = the sum of the clockwise moments about that point.
- All the forces acting on an object that don't pass through a fixed point can turn an object about that point.
- The direction of the force and the position of the fixed point determines whether the moment acts clockwise or anticlockwise.
- To calculate the force needed to stop an object turning we use the equation above. We need to know all the forces that don't act through the pivot, and the perpendicular distances from the forces' lines of action to the pivot.

Key word: principle of moments

On the up

To achieve the top grades, you should be able to carry out calculations to determine if an object with several forces acting on it is in equilibrium.

- If an object is in equilibrium it is balanced, not turning. We can take the moments about *any* point and will find that the total clockwise moment and the total anticlockwise moment about that point are equal. This is called the **principle of moments**.
- If we consider *any* point, a force acting on the object that has a line of action through that point will have a moment of zero. Any force with a line of action that does pass through that point has a non-zero moment, and has a turning effect on the object about that point.
- Whether the moment is clockwise or anticlockwise depends on the position of the point and the direction of the force.
- You can calculate the force needed to stop an object turning about a particular point using the principle of moments. You need to know all the forces that don't act through the pivot point, and the perpendicular distances from the forces' lines of action to the pivot. Add the moments to give the total moment. Supplying a moment with equal magnitude, and in the opposite direction, will stop the object turning.

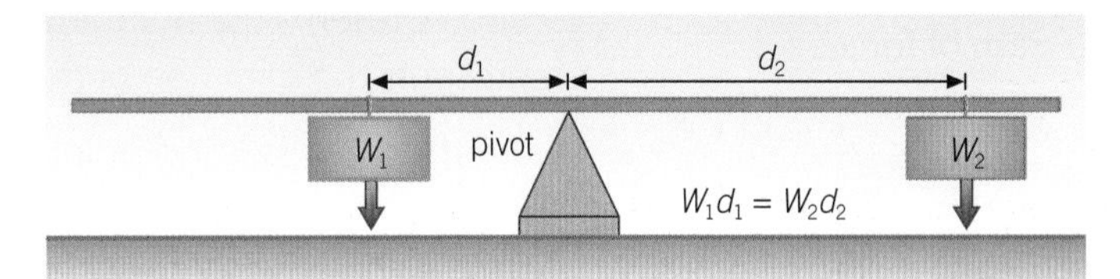

The principle of moments

- There are lots of everyday examples of the principle of moments, such as seesaws and balance scales.

The seesaw

1 If someone sits in the centre of a seesaw, the moment about the pivot is zero. Why?

2 A child sits on a seesaw. How can she increase her moment about the pivot?

Study tip

Practise doing moments calculations – sketch a diagram and tick off each force so you don't forget any.

8.8 The parallelogram of forces

Key points

- The parallelogram of forces is a scale diagram of two force vectors.
- The parallelogram of forces is used to find the resultant of two forces that do not act along the same line.
- You will need a protractor, a ruler, a sharp pencil, and a blank sheet of paper.
- The resultant is the diagonal of the parallelogram that starts at the origin of the two forces.

Key word: parallelogram of forces

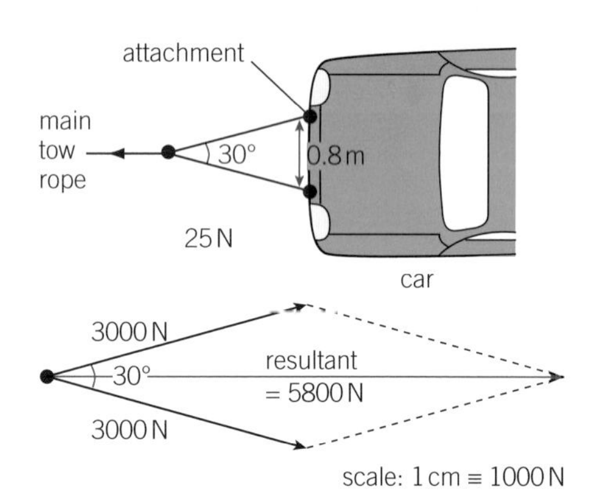

Using the parallelogram of forces

Study tip

Practise drawing parallelograms of forces. Always remember to state the scale of your diagram.

- To find the resultant of two forces that do not act along the same line, you cannot just add the forces, as you must take their directions into account.
- A **parallelogram of forces** is a scale diagram of two forces, which can be used to find their resultant.

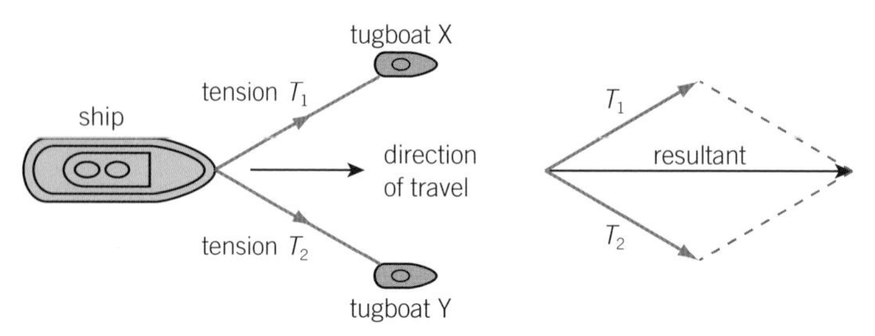

Combining forces

- To draw a parallelogram of forces:
  - Choose a scale for your diagram.
  - Draw an arrow to represent the first force. The length of the arrow, to scale, represents the magnitude of the force. The direction of the arrow represents the direction of the force.
  - From the same starting point as the first force, draw an arrow to represent the size and direction of the second force. The angle between the two arrows will be the same as the angle between the two forces.
  - The two arrows form adjacent sides of a parallelogram. A parallelogram has two pairs of parallel sides. Draw lines to complete the other two sides of the parallelogram.
  - Draw a line from the starting point of the two forces to the opposite corner of the parallelogram. This line represents the resultant of the two forces. The length of this diagonal line, to scale, represents the magnitude of the resultant force. The direction of this line represents the direction of the resultant force.
  - The direction of the resultant is usually expressed as an angle relative to one of the other forces.

1. An arrow is drawn to represent an 18 N force. The length of the arrow is 6.0 cm. What is the scale of the diagram?
2. A 3 N force and a 4 N force act at an angle of 45° to each other. Determine the size and direction of the resultant force by scale drawing.

8.9 Resolution of forces

Key points

- Resolving a force means finding perpendicular components that have a resultant force that is equal to the force.
- To resolve a force in two perpendicular directions, draw a rectangle with adjacent sides along the two directions so that the diagonal represents the force vector.
- For an object in equilibrium, the resultant force is zero.
- An object at rest is in equilibrium because the resultant force on it is zero.

On the up

To achieve the top grades, you should be able to resolve a force into two components acting at right angles to each other.

- In the same way that two forces can be combined to form a single resultant force, a single force can be split into two perpendicular components that together have the same effect as the single force. This is called resolving the force.
- To resolve a force into two perpendicular directions:
  - Choose a scale for your diagram.
  - Draw an arrow to represent the force. The length of the arrow, to scale, should represent the magnitude of the force. The direction of the arrow should represent the direction of the force.
  - From the same starting point, draw lines along the two perpendicular directions you want to resolve the vector into – these directions might be horizontal and vertical, or along a slope and perpendicular to the slope.
  - From the head of the arrow representing the force, draw lines along the same two perpendicular directions – this will form a rectangle with the force vector as its diagonal.
  - Measure the sides of the rectangle and use the scale to calculate the magnitude of the components.

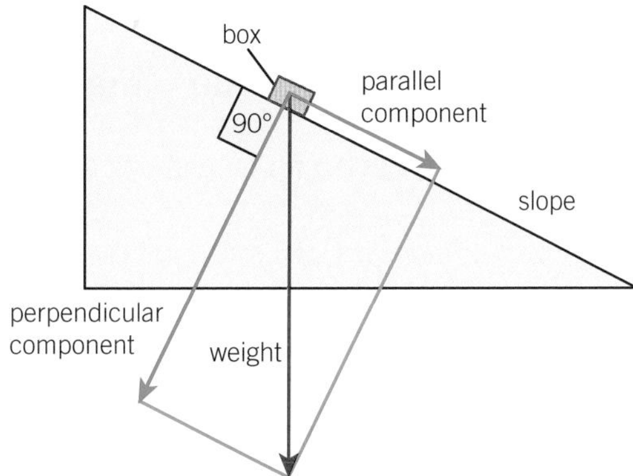

Resolving a force

- An object in equilibrium has no resultant force acting upon it, so it is either at rest or moving in a straight line at a constant speed.

1. What is meant by the components of a force?
2. Use a scale drawing to determine the horizontal and vertical components of a force of 13 N acting at 22.6° to the horizontal.

Summary questions

1. Give one example of a scalar quantity and one example of a vector quantity. [2 marks]
2. Identify what is meant by a contact force. [1 mark]
3. A child of weight 35 N sits 2.1 m from the centre of a seesaw.
   Calculate the moment of the child's weight about the seesaw pivot. [2 marks]
4. Describe where the centre of mass of a uniform metre rule is. [2 marks]
5. Define what is meant by a vector quantity. [1 mark]
6. Give Newton's third law of motion. [2 marks]
7. The resultant force on an object is zero.
   Describe the possible motion of the object. [3 marks]
8. Adam weighs 400 N and sits on a seesaw, 2.5 m from the pivot. Ben weighs 450 N and sits 2.0 m from the pivot, on the other side of the seesaw from Adam.
   Determine which child will move downwards. [3 marks]
9. Which of the following are scalar quantities and which are vector quantities? [4 marks]

   distance displacement speed velocity time mass weight acceleration
10. Explain how friction between the tyres and the road makes a car move forwards. [4 marks]
11. A force is applied at right angles to a lever, at 0.35 m from the pivot. The moment of the force is 4.2 N m.
    Calculate the magnitude of the force. [3 marks]
12. (H) A 3 N force and a 4 N force act at an angle of 90° to each other.
    Determine the resultant of the two forces by using a scale drawing. [4 marks]

Chapter checklist

Tick when you have:

reviewed it after your lesson	✓		
revised once – some questions right	✓	✓	
revised twice – all questions right	✓	✓	✓

Move on to another topic when you have all three ticks

8.1 Vectors and scalars	☐	☐	☐
8.2 Forces between objects	☐	☐	☐
8.3 Resultant forces	☐	☐	☐
8.4 Moments at work	☐	☐	☐
8.5 More about levers and gears	☐	☐	☐
8.6 Centre of mass	☐	☐	☐
8.7 Moments and equilibrium	☐	☐	☐
8.8 The parallelogram of forces	☐	☐	☐
8.9 Resolution of forces	☐	☐	☐

Student Book
pages 134–135

P9

9.1 Speed and distance–time graphs

Key points

- The speed of an object is:
  $v = \frac{s}{t}$
- The distance–time graph for any object that is:
  - stationary, is a horizontal line
  - moving at a constant speed, is a straight line that slopes upwards.
- The gradient of a distance–time graph for an object represents the object's speed.
- The speed equation $v = \frac{s}{t}$ can be rearranged to give:
  $s = v \times t$ or $t = \frac{s}{v}$

Key word: gradient

On the up

To get the top grades, you should be able to carry out calculations involving speed, distance, and time, which involve conversions to and from SI units.

- You can calculate the speed of an object using the equation $v = \frac{s}{t}$ where:
  v is the speed in metres per second, m/s
  s is the distance travelled in metres, m
  t is the time in seconds, s.
- You can rearrange this equation to give $s = v \times t$ or $t = \frac{s}{v}$

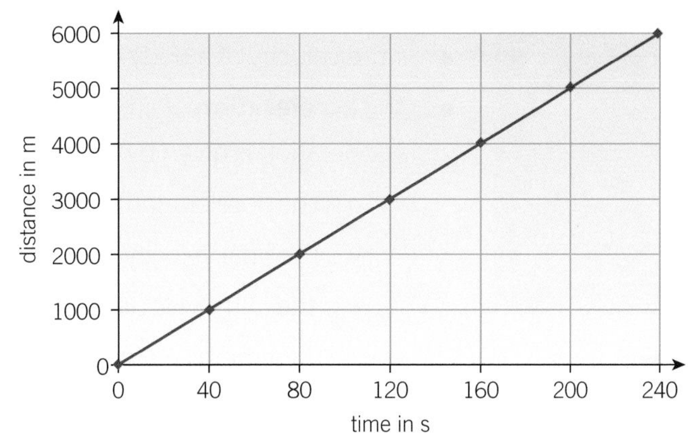

A distance–time graph

- A distance–time graph shows the distance of an object from a starting point (plotted on the y-axis) against the time taken (plotted on the x-axis).
  For a distance–time graph:
  - The **gradient** of the line on a distance–time graph represents speed, since speed is defined as the distance covered per unit of time.
  - If an object is stationary, the gradient is zero (the distance–time graph is a horizontal line).
  - If an object is moving at a constant speed, the distance–time graph is a straight line that slopes upwards.
  - The steeper the gradient, the greater the speed.

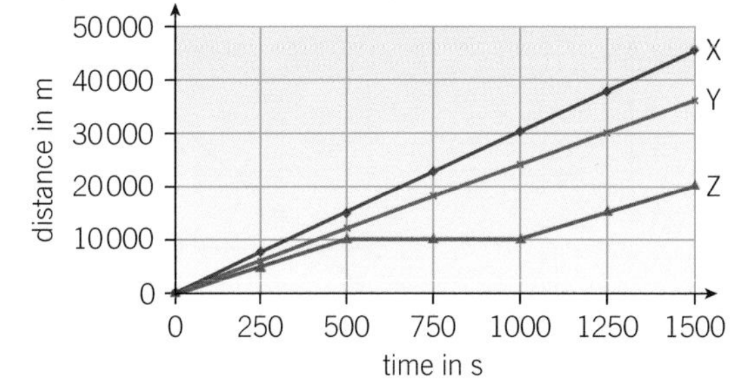

Comparing distance–time graphs

1 What is the speed of a runner who covers 400 m in 50 s?
2 What does the gradient of a distance–time graph represent?

Student Book pages 136–137 **P9**

9.2 Velocity and acceleration

Key points

- Velocity is speed in a given direction.
- A vector is a physical quantity that has a direction as well as a magnitude. A scalar is a physical quantity that has a magnitude only and does not have a direction.
- The acceleration of an object is $a = \frac{\Delta v}{t}$
- Deceleration is the change of velocity per second when an object slows down.

Synoptic links

For more information on acceleration, look at Topics P9.3 and P9.4.

For more information on vectors and scalars, look at Topic P8.1.

Key words: displacement, velocity, acceleration, deceleration

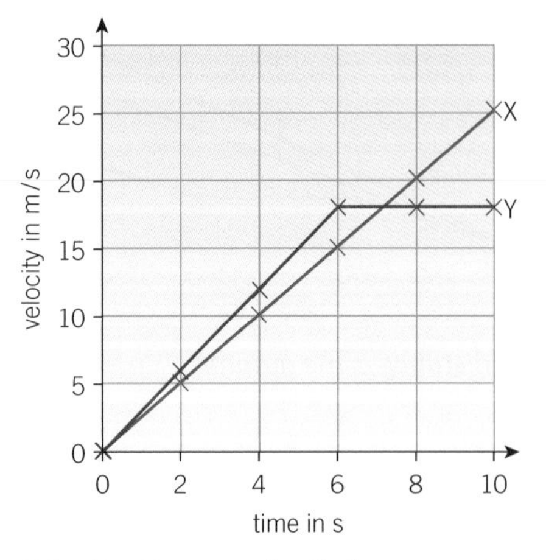

Velocity–time graphs. X shows constant acceleration; Y shows constant acceleration followed by constant speed

- The **displacement** of an object is the distance travelled in a given direction.
- The **velocity** of an object is its speed in a given direction.
- Scalar quantities have magnitude only, so speed and distance are both scalar quantities.
- Vector quantities have magnitude and an associated direction, so velocity and displacement are both vector quantities.
- If an object changes direction then it changes velocity, even if its speed stays the same.
- If the velocity of a body changes, we say that it accelerates.
- The **acceleration** of an object is its change in velocity per second. Acceleration is given by the equation $a = \frac{v - u}{t}$, where:

  a is the acceleration in metres per second squared, m/s^2

  v is the final velocity in metres per second, m/s

  u is the initial velocity in metres per second, m/s

  t is the time taken for the change in seconds, s
- This equation can also be written as $a = \frac{\Delta v}{t}$ where Δv represents the *change* in velocity, that is, $\Delta v = v - u$
- A **deceleration** is a negative acceleration – the object is slowing down.

An object moving in a circle at constant speed is continually changing direction, so it is velocity is continually changing. Hence, it is continually accelerating.

1 What is the difference between speed and velocity?

2 What is the S.I. unit of acceleration?

Student Book pages 138–139 **P9**

9.3 More about velocity–time graphs

Key points

- A motion sensor linked to a computer can be used to measure velocity changes.
- The gradient of the line on a velocity–time graph represents acceleration.
- If a velocity–time graph is a horizontal line, the acceleration is zero.
- A positive gradient on a velocity–time graph represents positive acceleration; a negative gradient represents deceleration.
- (H) The area under the line on a velocity–time graph represents distance travelled.

Synoptic link

For more information on motion graphs, look at Topics P9.1 and P9.4.

Study tip

Practise sketching velocity–time graphs for objects with different motion.

- You can use a motion sensor linked to a computer to record how the velocity of an object changes. You can collect the data and use it to plot a graph.
- A velocity–time graph shows the velocity of an object (plotted on the *y*-axis) against time taken (plotted on the *x*-axis).

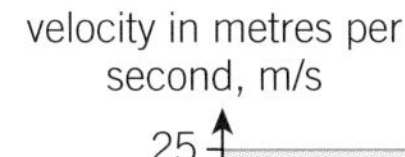

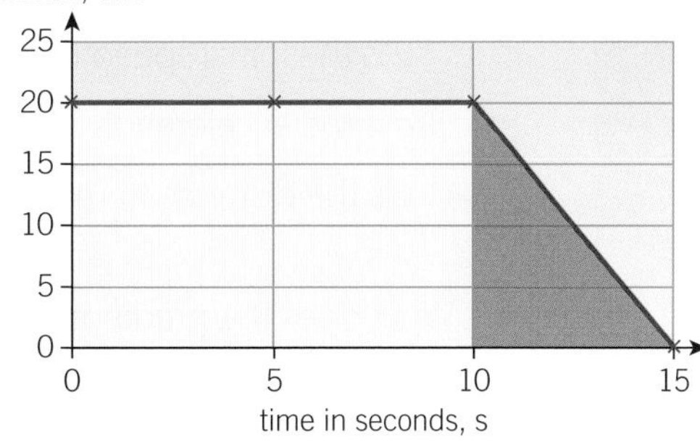

Velocity–time graph for a car

- The gradient of the line on a velocity–time graph represents acceleration, since acceleration is defined as the change in velocity per unit time.
- If the gradient of a velocity–time graph is zero (the graph is a horizontal line), the acceleration is zero, so the object is travelling at a steady speed.
- The steeper the gradient, the greater the acceleration.
- If the gradient of the line is negative, the object is decelerating.

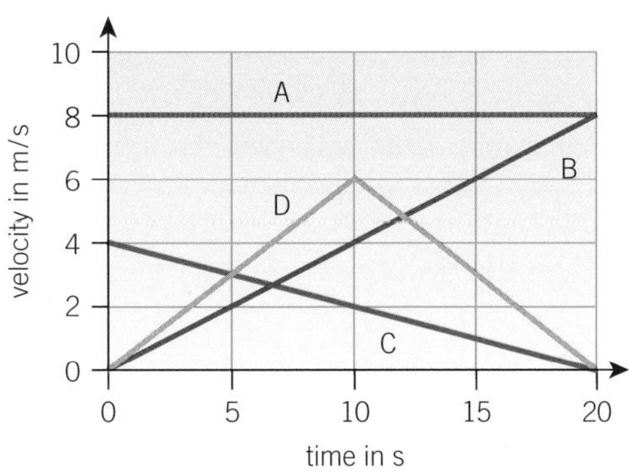

Velocity–time graph showing: A *constant velocity,* B *constant acceleration,* C *constant deceleration,* D *constant acceleration followed by constant deceleration*

Area under the line on a graph

Higher

The area under the line on a velocity–time graph represents the distance travelled in a given direction (or displacement). The bigger the area under the line, the greater the distance travelled.

1 Explain what a horizontal line represent on a velocity–time graph?

2 Describe the motion of the car in the velocity–time graph above.

Student Book pages 140–141

P9

9.4 Analysing motion graphs

Key points

- The speed of an object moving at constant speed is given by the gradient of the line on its distance–time graph.
- The acceleration of an object is given by the gradient of the line on its velocity–time graph.
- (H) The distance travelled by an object is given by the area under the line on its velocity–time graph.
- (H) The speed, at any instant in time, of an object moving at changing speed is given by the gradient of the tangent to the line on its distance–time graph.

Key word: tangent

Study tip

(H) Practise drawing tangents on curves at particular points. The tangent should just touch the curve at that point, but not cross it.

On the up

To achieve the top grades, you should be able to draw a tangent on a graph at a given point and calculate its gradient.

- You can determine an object's speed by calculating the gradient of its distance–time graph.
- You can determine an object's acceleration by calculating the gradient of its velocity–time graph.
- Distance, velocity, and acceleration are related by the equation $v^2 = u^2 + 2as$, where:

  v is the final speed in metres per second, m/s

  u is the initial speed in metres per second, m/s

  a is the acceleration in metres per second squared, m/s^2

  s is the distance in metres, m.

1 What does the gradient of the line on a velocity–time graph represent?

The distance–time graph for an object moving at a changing speed is a curve. The object's speed at a particular instant in time is given by drawing a **tangent** to the line at that instant, and calculating the gradient of the tangent.

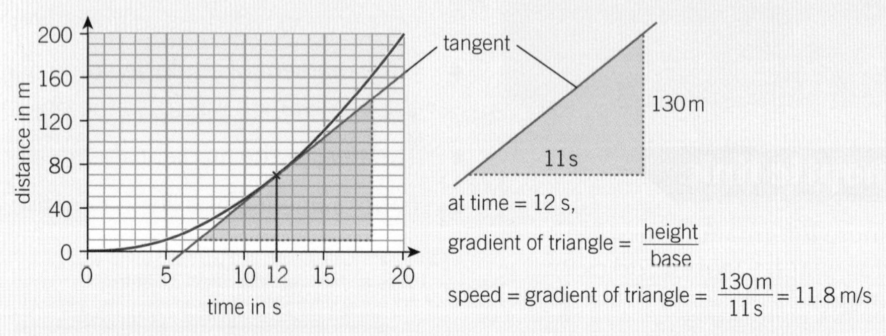

A distance–time graph for changing speed

Calculating the area under a velocity–time graph between two instants in time gives the distance travelled between those times.

2 What does an upwardly curving line represent on a distance–time graph?

3 What does an upwardly curving line represent on a velocity–time graph?

1. A car travels 1170 metres in 90 seconds. Calculate its speed. [2 marks]
2. The speed of a car increases by 20 m/s over 16 s. Calculate its acceleration. [2 marks]
3. Describe the motion of the object shown in **Figure 1**. [2 marks]
4. Define what the gradient of a velocity–time graph represents. [1 mark]
5. A vehicle travels 2400 m in 1 minute. Calculate its speed in m/s. [2 marks]
6. A car's speed increases from 15 m/s to 25 m/s in 5.0 s. Calculate the car's acceleration. [2 marks]

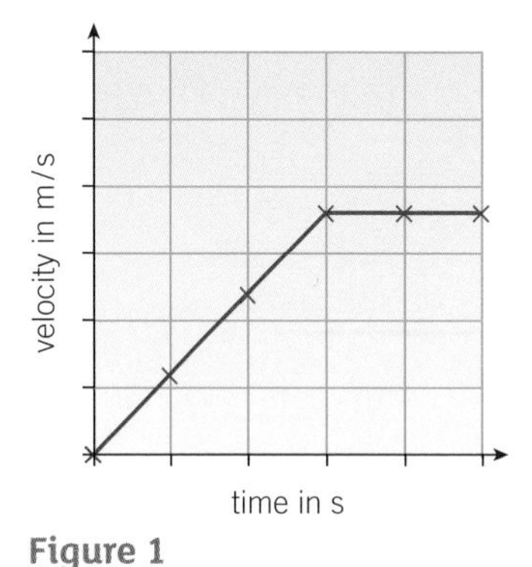

Figure 1

7. Describe the shape of a velocity–time graph for an object that travels at a constant speed in a straight line, then brakes to a stop with a constant deceleration. [3 marks]
8. **Figure 2** shows the velocity–time graph for part of a car journey over a 10 second period. Determine the acceleration of the car during this time. [3 marks]
9. A vehicle starts from rest and accelerates at $1.5\ m/s^2$ for 8.0 s. Calculate its final speed. [3 marks]
10. A vehicle travels 108 km in 1 hour. Calculate its speed in m/s. [3 marks]
11. (H) Determine the distance travelled by the car in **Figure 2** over the 10 second period. [4 marks]

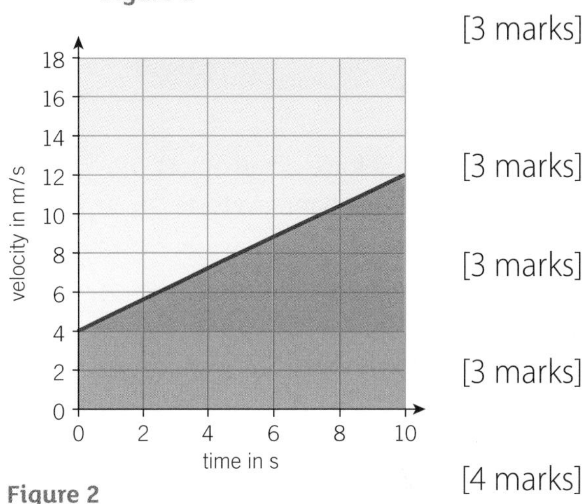

Figure 2

12. (H) **Figure 3** shows the distance–time graph for an object with a changing speed. Determine the speed of the object at a time of 15 seconds. [4 marks]

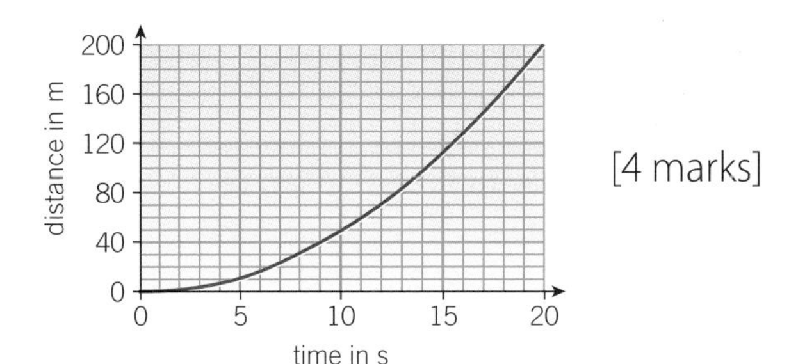

Figure 3

Chapter checklist

Tick when you have:

reviewed it after your lesson	✓		
revised once – some questions right	✓	✓	
revised twice – all questions right	✓	✓	✓

Move on to another topic when you have all three ticks

9.1 Speed and distance–time graphs
9.2 Velocity and acceleration
9.3 More about velocity–time graphs
9.4 Analysing motion graphs

Student Book pages 144–145 **P10**

10.1 Forces and acceleration

Key points

- The greater the resultant force on an object, the greater the object's acceleration.
- The greater the mass of an object, the smaller its acceleration for a given force.
- The resultant force acting on an object is $F = ma$
- (H) The inertia of an object is its tendency to stay at rest or in uniform motion.

1 The resultant force on an object increases. What happens to its acceleration?

2 What is the resultant force on a car of mass 1200 kg if its acceleration is 2.0 m/s²?

Synoptic link

For more information on resultant force, look at Topic P8.3.

Key words: Newton's second law of motion, inertia

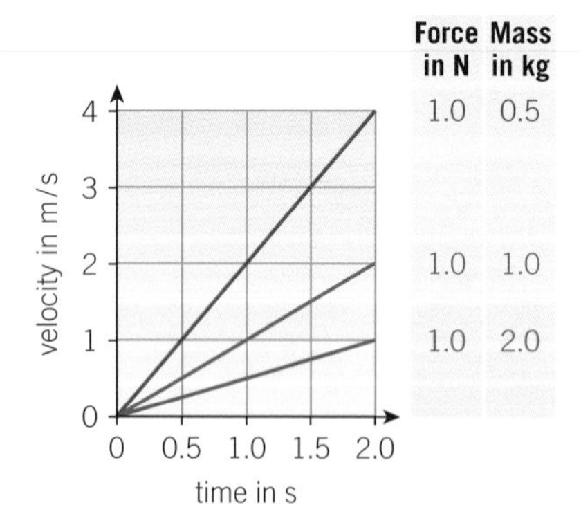

Velocity–time graph for different combinations of force and mass

- If a resultant force acts on an object, the object will accelerate in the direction of the resultant force.
- If an object is not accelerating, the resultant force on the object must be zero.
- Acceleration is a change in velocity. An object can change its velocity by changing its direction, even if it is travelling at a constant speed. So a resultant force is needed to make an object change direction.
- The resultant force acting on an object is given by the equation $F = m\,a$, where:

  F is the resultant force in newtons, N

  m is the mass of the object in kilograms, kg

  a is the acceleration of the object in metres per second squared, m/s²
- **Newton's second law of motion** says that the acceleration of an object is:
  - proportional to the resultant force on the object (that is, when there is a greater resultant force, the object experiences a greater acceleration)
  - inversely proportional to the mass of the object (that is, for a given force, a more massive object will experience a lesser acceleration).

The **inertia** of an object is its tendency to stay at rest or in uniform motion (moving at a constant speed in a straight line).

Investigating force and acceleration

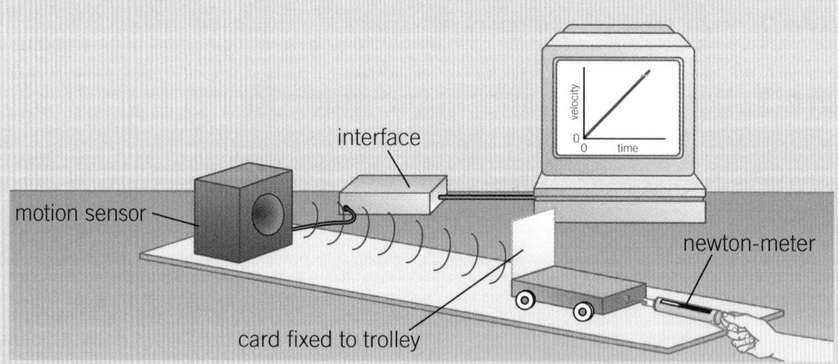

Investigating the link between force and acceleration

Investigate the acceleration of a trolley with constant mass by varying the force applied to it.

Use a motion sensor and computer to record the velocity of a trolley as it accelerates.

Use a newton-meter to pull the trolley a known distance with a constant force. You can repeat this and apply different forces to the trolley.

Investigate the acceleration of a trolley of varying mass whilst applying a constant force to it.

Keep the force the same and increase the mass of the trolley by standing small masses on it or by stacking trollies on top of each other.

Student Book pages 146–147 P10

10.2 Weight and terminal velocity

Key points

- The weight of an object is the force acting on the object due to gravity. Its mass is the quantity of matter in the object.
- An object acted on only by gravity accelerates at about 10 m/s^2.
- The terminal velocity of an object is the velocity it eventually reaches when it is falling. The weight of the object is then equal to the frictional force on the object.
- When an object is moving at terminal velocity, the resultant force on it is zero.

Synoptic link

For more information on gravitational field strength, look at Topic P1.4.

Key words: weight, mass, gravitational field strength, terminal velocity

Study tips

Write down all of the key words on this page and their corresponding units.

Remember that acceleration due to gravity and gravitational field strength have different units.

- The **mass** of an object is the quantity of matter in it.
- If an object falls freely, the resultant force acting on it is the force of gravity. This force will make the object accelerate at about 10 m/s^2 towards the Earth's surface.
- An object's **weight** is the force of gravity that acts on the object.
- The acceleration, which the object experiences due to the force of gravity, is referred to as the acceleration due to gravity.
- When the force acting on an object is its weight, the equation $F = m\,a$ becomes $W = m\,g$, where:

  W is the weight in newtons, N

  g is the acceleration due to gravity in in metres per second squared, m/s^2.
- The gravitational force on a 1 kg object is called the **gravitational field strength** at the place where the object is. Its units are newtons per kilogram, N/kg.
- When an object falls through a fluid (liquid or gas), the fluid exerts a drag force on the object, resisting its motion. The faster the object falls, the bigger the drag force becomes, until eventually it will be equal to the weight of the object. The resultant force is now zero, so the object stops accelerating and moves at a constant velocity called the **terminal velocity**.

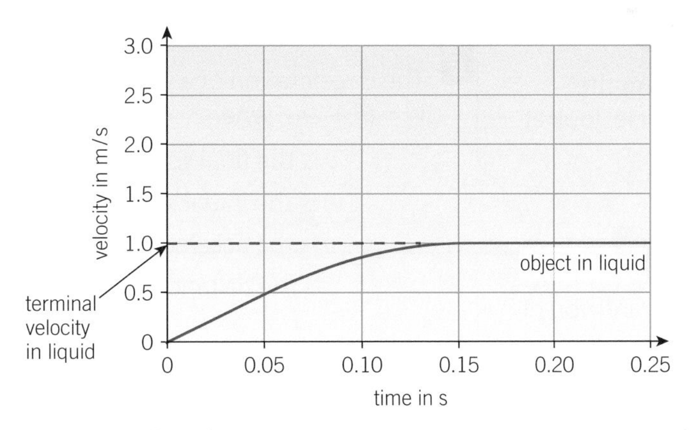

Falling in a liquid

- For an object falling through air, the drag force is called air resistance.

1 Why does an object dropped in a fluid initially accelerate?

2 What eventually happens to the acceleration of an object falling through a fluid?

On the up

To achieve the top grades, you should be able to explain the motion of an object falling through a fluid in terms of the forces acting on the object.

Student Book **pages 148–149** P10

10.3 Forces and braking

Key points

- Friction and air resistance oppose the driving force of a vehicle.
- The stopping distance of a vehicle depends on the thinking distance and the braking distance.
- High speed, poor weather conditions, and poor vehicle maintenance all increase the braking distance. Poor reaction time (due to tiredness, alcohol, drugs, or using a mobile phone) and high speed both increase the thinking distance.
- $F = m\,a$ gives the braking force of a vehicle.

- If a vehicle is travelling at a steady speed the resultant force on it is zero. The driving forces are equal and opposite to the frictional forces (which are air resistance and friction between the tyres and the road).
- The faster the speed of a vehicle, the bigger the deceleration needed to bring it to rest within a particular distance. A greater braking force is needed to produce a greater deceleration.
- Also, the greater the mass, the greater the braking force needed to produce a given deceleration.
- The **stopping distance** of a vehicle is the distance it travels during the driver's reaction time (the **thinking distance**) plus the distance it travels under the braking force (the **braking distance**).
- The thinking distance is increased if the driver is tired, under the influence of alcohol or drugs, or using a mobile phone.
- The braking distance can be increased by poorly maintained roads, bad weather conditions, or poorly maintained vehicles (for example, worn tyres or worn brakes will increase braking distance).

1. What is the resultant force on a car travelling at a steady speed on a straight horizontal road?
2. What is the relationship between stopping distance, thinking distance, and braking distance?

Synoptic link

For more on calculating the deceleration of a vehicle, look at Topic P9.4.

Study tip

Draw a diagram to remind you of the relationship between thinking distance, braking distance, and stopping distance.

Higher

Deceleration

The deceleration of a vehicle can be calculated using the equation $v^2 = u^2 + 2as$, where:

v is the final speed in metres per second, m/s

u is the initial speed in metres per second, m/s

a is the acceleration in metres per second squared, m/s^2

s is the distance in m.

This can be rearranged to give $s = -\frac{u^2}{2a}$ and $a = -\frac{u^2}{2s}$

Key words: stopping distance, thinking distance, braking distance

Higher tier

Student Book **pages 150–151** P10

10.4 Momentum

Key points

- The momentum of a moving object is $p = m\,v$
- The unit of momentum is kg m/s.

- All moving objects have **momentum**. The greater the mass and the greater the velocity of an object, the greater its momentum.
- Momentum has both a size and a direction, so it is a vector quantity.
- You can calculate momentum using the equation $p = m\,v$, where:

  p is the momentum in kilograms metres per second, kg m/s

  m is the mass in kilograms, kg

  v is the velocity in metres per second, m/s.

- A closed system is a system in which the total momentum before an event is the same as the total momentum after the event. This is called conservation of momentum.

Key words: momentum, conservation of momentum

- A closed system is a system of objects that has no resultant force acting on it.
- The law of **conservation of momentum** states that whenever objects interact in a closed system, the total momentum before the interaction is equal to the total momentum after the interaction. Another way to say this is that the overall change in momentum within the closed system is zero.
- Momentum is conserved in a collision or an explosion as no external forces act on the objects. After a collision, the colliding objects may move off together or they may move apart.

1 What is the momentum of a stationary object?
2 What is the momentum of a 1500 kg car travelling at 20 m/s?
3 What does the law of conservation of momentum state?

Student Book pages 152–153 P10

10.5 Using conservation of momentum

Key points

- Momentum is defined as mass × velocity, and has both size and direction.
- When two objects push each other apart, they move with different speeds if they have unequal masses, and with equal and opposite momentum, so their total momentum is zero.
- Use the equation $m_A v_A + m_B v_B = 0$ when two objects, **A** and **B**, recoil from each other.

Study tip

When practising conservation of momentum calculations, sketch a diagram to show the motion of the objects before and after the collision. Mark on your diagram which direction is positive.

- Momentum has both magnitude and direction, so it is a vector quantity.
- In calculations, one direction must be defined as positive, so momentum in the opposite direction is negative.
- When two objects are at rest their momentum is zero. In an 'explosion' the objects move apart with equal and opposite momentum. One object has momentum in the positive direction and the other in the negative direction, so the total momentum after the explosion is still zero.
- For two objects, **A** and **B**, that are initially stationary:
  $m_A v_A + m_B v_B = 0$, where:
  $m_A v_A$ is the momentum of **A** after the explosion, and
  $m_B v_B$ is the momentum of **B** after the explosion.
- Firing a bullet from a gun is an example of an explosion. The bullet moves off with a momentum in one direction and the gun 'recoils' with an equal momentum in the opposite direction.

1 What is the total momentum after an explosion equal to?
2 Two students on roller skates stand still, holding each other, in the playground. They push each other away.
What can you say about the momentum of each student?
3 What is the momentum of a car of mass 1200 kg travelling at 30 m/s?

10.6 Impact forces

Key points

- When vehicles collide, the force of the impact depends on mass, change of velocity, and the length of the impact time.
- The longer the impact time, the more the impact force is reduced.
- When two vehicles collide:
  - they exert equal and opposite forces on each other
  - their total momentum is unchanged.
- Impact force = change of momentum ÷ impact time, so the shorter the impact time, the greater the impact force.

- The impact force that an object exerts in a collision is given by the equation $F = \frac{m\Delta v}{t}$, where:

  m is the mass of the object in kilograms, kg

  Δv is the change in velocity in metres per second, m/s

  t is the time taken for the collision to occur in seconds, s.
- From this equation, you can see that:
  - the greater the mass, the greater the impact force
  - the greater the change in velocity, the greater the impact force
  - the shorter the time taken for the collision, the lesser the impact force.
- Since

  change in momentum = mass × change in velocity

  this equation can also be written as

  $$\text{impact force} = \frac{\text{change in momentum}}{\text{time}}$$
- When two vehicles collide:
  - they exert equal and opposite forces on each other
  - the change in momentum of one vehicle is equal and opposite to the change in momentum of the other vehicle
  - so their total momentum remains unchanged.

A crash test

- Crumple zones in cars are designed to fold in a collision. These increase the impact time during a collision and so reduce the force on the car and the people in it.

1 What is a crumple zone designed to do?

2 In a collision, what does the size of the impact force depend on?

10.7 Safety first

Key points

- Cycle helmets and cushioned surfaces (e.g., in playgrounds) reduce impact forces by increasing the impact time.
- Seat belts and air bags spread the force across the chest and increase the impact time.
- Side impact bars and crumple zones give way in an impact, and so increase the impact time.
- Conservation of momentum can be used to find the speed of a car before an impact.

Study tip

Draw a labelled diagram of a car, showing the safety features.

On the up

You should be able to explain safety features with reference to collisions and the rate of change of momentum.

- Cycle helmets crumple on impact and reduce the impact force by increasing the impact time. Cushioned playground surfaces increase the impact time of a falling child and so also decrease the impact force on the child.
- Modern cars contain a number of safety features designed to reduce the forces on the driver and passengers in a collision.
    - Seat belts and airbags spread the impact force on the head and body across a larger area.
    - Airbags increase the impact time on the driver's head and chest (which would be much shorter if the driver were to hit the steering wheel) and so decrease the impact force.
    - A seat belt stops a person being flung forward if the car stops suddenly. The seatbelt stretches slightly; increasing the impact time and reducing the force.
    - Side impact bars and crumple zones fold up in a collision, which increases the impact time and reduces the impact force.

An airbag in action

rear crumple zone
airbags
front crumple zone
rear seat belts
side impact bars
front seat belts
collapsible steering wheel

Car safety features

- Because momentum is conserved in a collision, police can use the masses of the vehicles involved in a collision, along with data for speeds and braking distances, to calculate the speed of a car before the collision.

1 What would happen to a car passenger in a head-on collision if they were not wearing a seat belt?

2 What would happen to a car passenger in a head-on collision if they were wearing a very narrow seat belt?

Student Book pages 158–159

10.8 Forces and elasticity

Key points

- An object is described as elastic if it returns to its original shape after removing the force deforming it.
- The extension is the difference between the length of the object and its original length.
- The extension of a spring is directly proportional to the force applied to it, as long as the limit of proportionality is not exceeded. This relationship is linear.
- Beyond the limit of proportionality, the extension of a spring is no longer proportional to the applied force. This relationship becomes non-linear.

- An **elastic** object is one that regains its shape when the forces deforming it are removed.
- If you hang small weights from a spring it will stretch. The increase in the spring's length from the original length is called the **extension**.
- If you plot a graph of a spring's extension against force applied to it, you get a straight line that passes through the origin. This tells you that the extension is **directly proportional** to the force applied. However, if you apply too big a force to the spring, the line begins to curve because you have exceeded the **limit of proportionality**.
- Objects and materials that behave like this are said to obey Hooke's law, which states that extension is directly proportional to the force applied, provided the limit of proportionality is not exceeded.
- You can express Hooke's law as an equation, $F = k\,e$, where:

  F is the force applied in newtons, N

  k is the spring constant of the spring in newtons per metre, N/m

  e is the extension in metres, m.
- The stiffer a spring is, the greater its spring constant.
- When an elastic object is stretched, work is done and the elastic potential energy store of the object increases.
- Hooke's law also applies to an elastic object when it is compressed.

1 What is an inelastic object?

2 How can you tell when a spring has exceeded its limit of proportionality?

Synoptic link

For more information about elastic potential energy, look at Topic P1.5.

Key words: elastic, extension, directly proportional, limit of proportionality

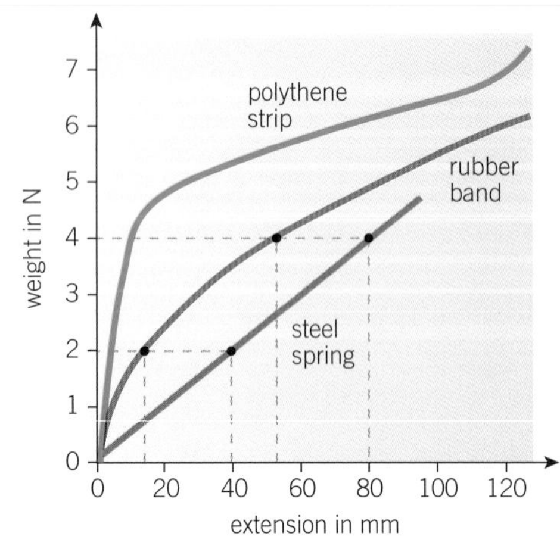

Extension against force for different materials

Stretch tests

Clamp a spring at its upper end and hang an empty weight hanger from the other end to keep the spring straight.

Measure the original length of the spring with a metre rule.

Add weights to the bottom of the spring and measure the new length.

For each weight you add, subtract the spring's original length from the stretched length to give the extension of the spring.

Keep adding more weights and calculate the extension each time. Record the force (weight) and extension each time.

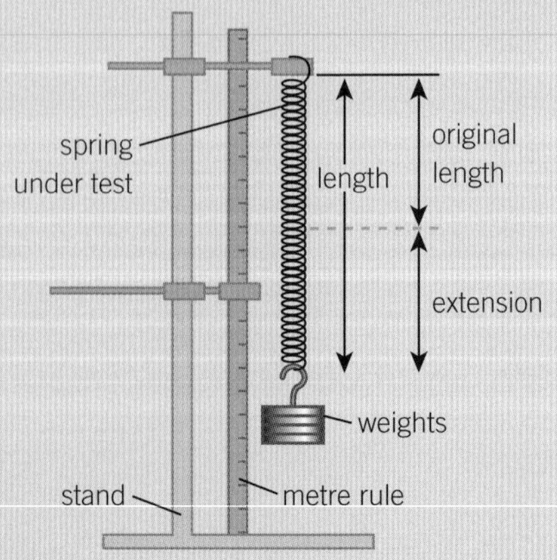

Plot a graph of force against extension for the spring.

spring under test
length
original length
extension
weights
stand
metre rule

Investigating stretching

1. Calculate the resultant force on an object of mass 45 kg that is accelerating at 11 m/s^2. [2 marks]
2. Calculate the weight of a student of mass 55 kg. Gravitational field strength = 9.8 N/kg. [2 marks]
3. Name two things that would increase the braking distance of a car. [2 marks]
4. Describe the graph of force against extension for a spring, up to its limit of proportionality. [2 marks]
5. Calculate the momentum of a 180 g toy car moving at a speed of 4 m/s. [2 marks]
6. Copy the axes in **Figure 1**, and sketch the force–extension graph for a spring up to its limit of proportionality. [2 marks]

force
extension

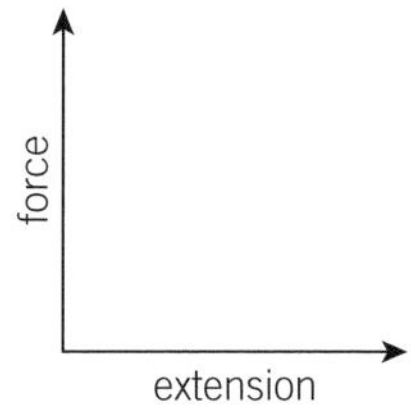

Figure 1

7. (H) A car of mass 900 kg is travelling at a speed of 20 m/s when the brakes are applied. It is brought to rest in 4.0 s. Calculate the braking force. [3 marks]
8. (H) Explain why a horse rider who falls off is less likely to have a bad head injury if she is wearing a safety helmet. [3 marks]
9. (H) Describe what is meant by the inertia of an object. [1 mark]
10. (H) A bullet of mass 10 g is shot from a gun of mass 3.8 kg with a velocity of 860 m/s. Calculate the recoil speed of the gun. [3 marks]
11. (H) A car in a crash test hits an obstacle and decelerates to rest in 0.75 s. The mass of the car is 900 kg and the resultant force on the car is 32 000 N. Calculate the initial speed of the car. [4 marks]
12. (H) Explain how landing on a soft, cushioning crash mat after performing a vault helps a gymnast avoid injury. [4 marks]

Chapter checklist

Tick when you have:

reviewed it after your lesson ✓ ☐ ☐

revised once – some questions right ✓ ✓ ☐

revised twice – all questions right ✓ ✓ ✓

Move on to another topic when you have all three ticks

Topic			
10.1 Force and acceleration	☐	☐	☐
10.2 Weight and terminal velocity	☐	☐	☐
10.3 Forces and braking	☐	☐	☐
10.4 Momentum	☐	☐	☐
10.5 Using conservation of momentum	☐	☐	☐
10.6 Impact forces	☐	☐	☐
10.7 Safety first	☐	☐	☐
10.8 Forces and elasticity	☐	☐	☐

Student Book **pages 162–163** **P11**

11.1 Pressure and surfaces

Key points

- Pressure is the force normal to a surface ÷ area of the surface.
- Pressure $p = \frac{F}{A}$
- The unit of pressure is the pascal (Pa), which is equal to 1 N/m^2.
- The force F or area A can be calculated by rearranging the pressure equation $p = \frac{F}{A}$ to give $F = pA$ or $A = \frac{F}{p}$

- When a force acts at right angles to a surface, the force exerts pressure on the surface. Pressure is the force exerted per unit area.
- Pressure is measured in pascals. 1 pascal is equal to a force of 1 N acting over a surface area of 1 m^2.
- You can calculate pressure on a surface using the equation $p = \frac{F}{A}$ where:

  p is the pressure in pascals, Pa (or N/m^2)

  F is the force in newtons, N

  A is the cross-sectional area at right angles (normal) to the direction of the force in metres squared, m^2.
- You can rearrange this equation to give $F = pA$ or $A = \frac{F}{p}$

Pressures often have large values so may be expressed as prefixes or powers of ten, for example, 1000 Pa = 10^3 Pa = 1 kPa and 1000 000 Pa = 10^6 Pa = 1 MPa.

1. What is the pressure exerted on the ground by a person of weight 300 N if the area of their feet in contact with the ground is 0.04 m^2?
2. Why is a person unlikely to damage a wooden floor when wearing trainers, but very likely to damage it when wearing stiletto heels?

Higher tier

Student Book **pages 164–165** **P11**

11.2 Pressure in a liquid at rest

Key points

- The pressure in a liquid increases with increasing liquid depth.
- A liquid flows until the pressure along the same horizontal line is constant.
- The greater the density of a liquid, the greater the pressure in the liquid.
- The pressure p due to the column height h of liquid of density ρ is given by the equation $p = h \rho g$

- A liquid will flow until the pressure along any horizontal line in the liquid is constant.
- The pressure at a particular point in a liquid is calculated using the equation $p = h \rho g$, where:

  p is the pressure in pascals, Pa

  h is the height of liquid above that point in metres, m

  ρ is the density of the liquid in kilograms per cubic metre, kg/m^3

  g is the gravitational field strength in newtons per kilogram, N/kg.
- The further down you go in a liquid, the greater the height of liquid above you, so the greater the pressure of the liquid.
- The greater the density of a liquid, the greater the pressure in the liquid.

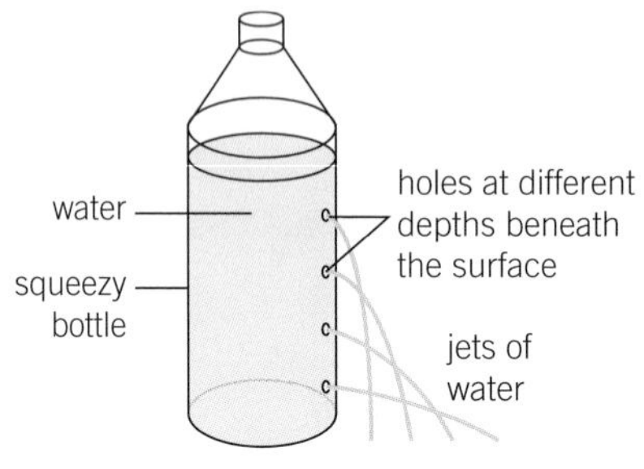

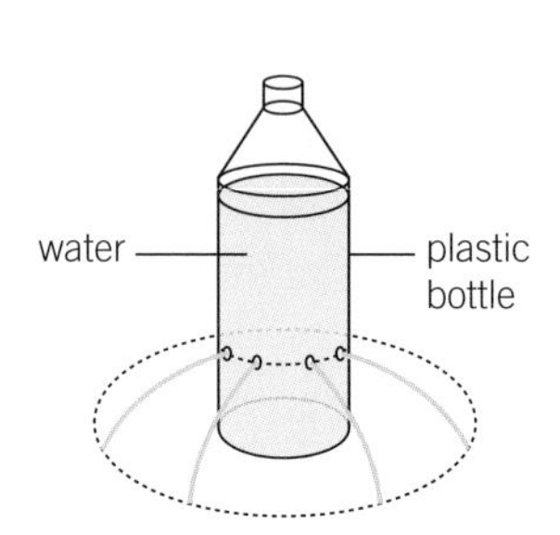

Pressure in a liquid at rest

Synoptic link

For more information on forces in liquids, look at Topic P11.4.

1 What is the pressure due to a column of water of height 0.5 m, and density 1000 kg/m^3?

2 Pressure is usually expressed in pascals, Pa. Name another unit for pressure.

Student Book **pages 166–167** **P11**

11.3 Atmospheric pressure

Key points

- Air molecules collide with surfaces and create pressure on them.
- Atmospheric pressure decreases with higher altitude because there is less air above a given altitude than there is at a lower altitude.
- The density of the atmosphere decreases with increasing altitude.
- The force on a flat object due to a pressure difference = the pressure difference × the area of the flat surface.

On the up

You should be able to explain why atmospheric pressure varies with height above a surface.

- The atmosphere is a layer of air around the Earth.
- Atmospheric pressure is due to air molecules within the Earth's atmosphere colliding with surfaces. Every impact of an air molecule with the surface creates a tiny force on the surface, and pressure is the force per unit area.
- The higher the altitude, the less weight of air there is above that altitude, so the lower the pressure.
- The higher the altitude, the less dense the air is as there are fewer molecules, so the lower the pressure.

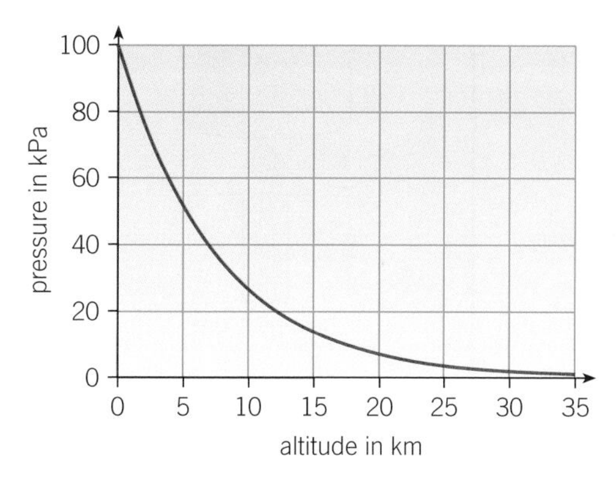

Graph of atmospheric pressure against altitude

- Flat objects (e.g., windows) with a different pressure on either side will experience a force because of the pressure difference. This force can be calculated using the equation:
  force = pressure difference × the area of flat surface

1 What is the SI unit of atmospheric pressure?

2 A climber descends from the top of a tall mountain to sea level. What happens to the atmospheric pressure acting on him?

11.4 Upthrust and flotation

Key points

- The upthrust on an object in a fluid:
  - is an upward force on the object due to the fluid
  - is caused by the pressure within the fluid.
- The pressure at a point in a fluid depends on the density of the fluid and the depth of the fluid at that point.
- An object sinks if its weight is greater than the upthrust on it when it is fully immersed.

Key word: upthrust

Study tip

Draw a labelled diagram to summarise all the information on this page about upthrust and floating.

On the up

To achieve the highest grades, you should be able to describe the factors which influence whether an object will float or sink.

- When an object is immersed in a fluid it displaces a volume of fluid equal to its own volume.
  - Pressure in a fluid increases with depth, so the pressure exerted by the fluid on the bottom of the object is greater than the pressure exerted on the top of the object. This means that the upward force of the water on the bottom of the object is greater than the downward force of the water on the top of the object. The resultant of these two forces is called the **upthrust**.
  - The upthrust is equal to the weight of fluid displaced.

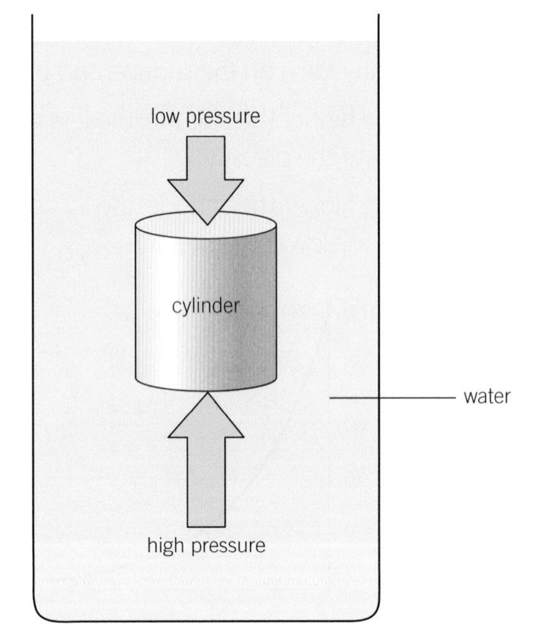

Explaining upthrust

- An object floats when its weight is equal to the upthrust. An object sinks when its weight is greater than the upthrust.
  - An object that is less dense than the liquid it is immersed in will float.
  - An object that is more dense than the liquid it is in will sink.

1 What is the relationship between pressure and depth in a fluid?
2 What is the name of the upward force on an object immersed in a fluid?

1. A paving slab exerts a force of 150 N when laid with an area of 0.25 m^2 in contact with the ground.
   Calculate the pressure exerted by the paving slab on the ground. [2 marks]
2. Describe how atmospheric pressure changes with height above sea level. [1 mark]
3. Explain why atmospheric pressure changes with height above sea level. [3 marks]
4. Explain how a suction cap works. [3 marks]
5. A woman of mass 65 kg stands on the floor with bare feet. The total area of her feet in contact with the floor is 0.020 m^2.
   Calculate the pressure she exerts on the floor. [3 marks]
6. (H) Calculate the pressure at the bottom of a pond 0.75 m deep, due to the water in the pond. The density of the water is 1000 kg/m^3. [3 marks]
7. (H) Calculate the pressure due to a column of mercury 75 cm high. The density of mercury is 14 000 kg/m^3. [3 marks]
8. (H) Describe a demonstration to show that the pressure in a liquid is the same at the same depth. [3 marks]
9. (H) A vehicle of weight 24 000 N exerts a pressure of 8 kN on the ground. The vehicle is fitted with caterpillar tracks.
   Calculate the area of the tracks in contact with the ground. [3 marks]
10. (H) The pressure at a particular point on the seabed, due to the seawater above, is 4.2×10^6 Pa. Calculate the depth of the water at this point. The density of sea water is 1050 kg/m^3. Gravitational field strength = 9.8 N/kg. [4 marks]
11. (H) Explain how upthrust comes about when an object is immersed in water. [4 marks]
12. (H) A metal object is weighed using a newton-meter. The object is then lowered into a beaker of water until it is fully immersed.
    Explain what happens to the reading on the newton-meter. [3 marks]

Chapter checklist ✓

Tick when you have:

reviewed it after your lesson	✓		
revised once – some questions right	✓	✓	
revised twice – all questions right	✓	✓	✓

Move on to another topic when you have all three ticks

11.1 Pressure and surfaces	☐	☐	☐
11.2 Pressure in a liquid at rest	☐	☐	☐
11.3 Atmospheric pressure	☐	☐	☐
11.4 Upthrust and flotation	☐	☐	☐

Practice questions

01 **Figure 1** shows a vase.
The mass of the vase is 1.3 kg.

01.1 Calculate the weight of the vase. ($g = 9.8$ N/kg) [3 marks]

01.2 Which of the following is the unit of weight? Choose the correct answer from the list below.

joule kilogram newton [1 mark]

01.3 Sketch the diagram and put a cross on the diagram to show a possible position of the centre of mass of the vase. [1 mark]

Figure 1

02 The distance–time graph for a car journey is shown in **Figure 2**.

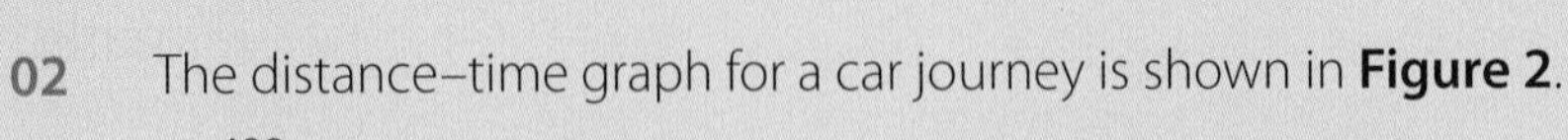

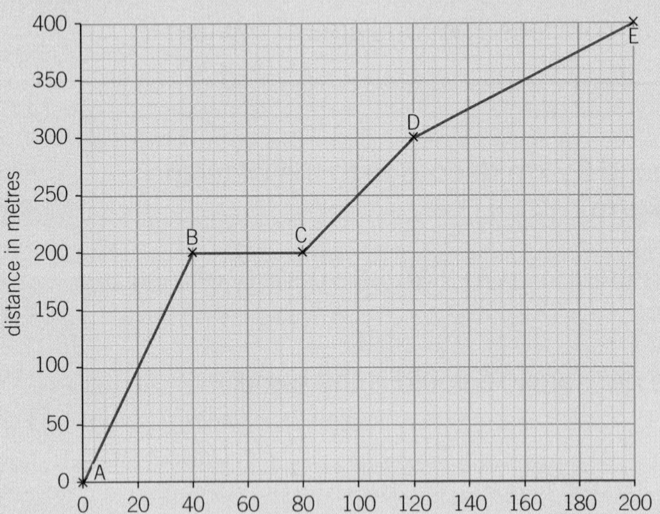

Figure 2

02.1 The car travelled fastest between points A and B. Explain how the graph shows this. [2 marks]

02.2 The car was stopped at traffic lights during part of its journey. For how long was the car stopped at the traffic lights? [2 marks]

02.3 Describe the motion of the car between points D and E. [1 mark]

02.4 A motorcycle started the same journey at the same time as the car. The motorcycle travelled at a constant speed and completed the journey in 160 seconds. Describe how you could add a second line on the same axes to represent the journey of the motorcycle. [2 marks]

03 Some students are investigating the speed at which cars travel along a straight road close to their school.
The students stand together at the side of the road. One student starts a stopwatch as a car passes. He stops the stopwatch when he sees the car pass a tree about fifty metres down the road. The students use a metre rule, and repeatedly measure 1 m distances, to measure the total distance from their start point to the tree.

03.1 Describe what the students should do to improve the accuracy of their measurements of distance and time. [5 marks]

Study tip

Think about how the students should make the measurements and the measuring instruments they should use.

03.2 Give three things the students could do to make their investigation as safe as possible. [3 marks]

03.3 Give the equation the students should use to calculate the speed of a car from their measurements. [1 mark]

03.4 The students carry out their investigation on a Wednesday morning. They calculate the speed of each car. They record how many cars are breaking the speed limit. Their results are shown in **Table 1**.
Explain how you would expect the students' results to be different if they carried out their investigation at the same place and time, but on a Sunday morning. [4 marks]

Table 1

Time of day	Number of cars breaking the speed limit
7:31 – 8:00	12
8:01 – 8:30	3
8:31 – 9:00	1
9:01 – 9:30	4
9:31 – 10:00	10
10:01 – 10:30	11

04 A student investigates how the extension of a spring varies with the force applied to it. The student uses the spring, a set of masses, and a metre rule. He suspends the spring from a laboratory stand and measures its length. He then hangs a mass from it and measures its new length. He subtracts the original length from the new length to give the extension of the spring.
He adds more masses to the spring and, each time, subtracts the original length from the new length to obtain a set of values of mass against extension for the spring.

04.1 What is the independent variable in this investigation? [1 mark]

04.2 The force applied to the spring is the weight of the mass hanging on the spring. Write down the equation the student should use to calculate the weight of the mass. [1 mark]

04.3 The graph in **Figure 3** shows the student's results.
Determine the spring constant of the spring. [3 marks]

04.4 Estimate the extension of the spring at the limit of proportionality. Give a reason for your answer. [2 marks]

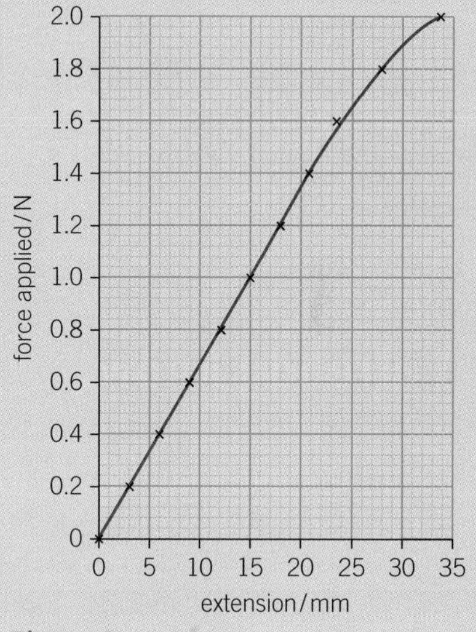

Figure 3

05 A skydiver jumps from a plane. The skydiver is shown in **Figure 4**. Arrows X and Y represent two forces acting on the skydiver as she falls through the air. Force Y is the weight of the skydiver, and gravitational field strength = 9.8 N/kg.

Figure 4

05.1 The weight of the skydiver is 686 N. Calculate the mass of the skydiver. [3 marks]

05.2 What causes force X? [1 mark]

05.3 The graph in **Figure 5** shows how the velocity of the skydiver varies with time as the skydiver falls.
Explain how the velocity of the skydiver varies with time. You answer should include references to the graph and the sizes of forces X and Y. [6 marks]

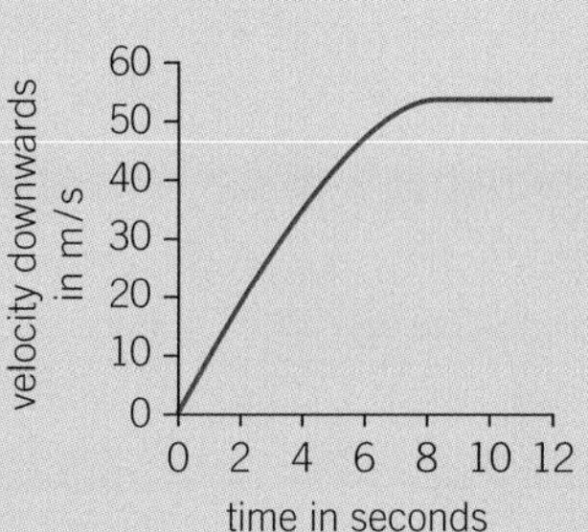

Figure 5

06 A resultant force acts on an object.

06.1 What is meant by a resultant force? [2 marks]

06.2 (H) A 30 N force and a 40 N force act on an object at an angle of 45° to each other. Determine the magnitude and the direction of the resultant force on the object by scale drawing. [4 marks]

Study tip

Always have a protractor, ruler, and sharp pencil for drawing accurate scale diagrams. Make sure you state what scale you are using.

Waves, electromagnetism, and space

When you speak into a mobile phone, you create sound waves that carry information. These waves are detected by a microphone that produces electrical waves in the phone circuits. Your phone then sends out radio waves that carry the information to your mobile phone network and then to the person you are calling. Medical doctors use radio waves in scanners to obtain 3D images of organs. They also use X-rays and ultrasonic waves to visualise objects inside the body.

In this section, you will learn about waves and their properties, and the many ways they are used. You will also learn about magnetic fields and how we use them to produce electrical waves.

I already know...	I will revise...
the top of a water wave is called a crest and the bottom is called a trough.	how the wavelength of a wave depends on its speed and its frequency.
light travels much faster than sound and can travel through space whereas sound cannot.	how to measure the speed of sound waves in air and in a solid.
the spectrum of white light is continuous from red to orange, yellow to green, and blue to violet.	how electromagnetic waves carry information and how they are used to form images.
there are different types of waves, such as sound waves and electromagnetic waves, but they all have common properties such as refraction.	what we mean by refraction of waves when they cross a boundary between different substances.
a magnet lines up with the Earth's magnetic field.	how the strength of a magnetic field is measured and what a solenoid is.
an electric motor is used to turn objects. An electric generator produces an electric current when it turns.	how an electric motor and an electric generator work.
satellites orbit the Earth.	how satellites stay in their orbits and what we mean by a geostationary satellite.

Student Book
pages 174–175

P12

12.1 The nature of waves

Key points

- Waves can be used to transfer energy and information.
- Transverse waves oscillate perpendicular to the direction of energy transfer of the waves. Ripples on the surface of water are transverse waves. So are all electromagnetic waves.
- Longitudinal waves oscillate parallel to the direction of energy transfer of the waves. Sound waves in air are longitudinal waves.
- Mechanical waves need a medium (a substance) to travel through. They can be transverse or longitudinal waves.

Study tip

Draw labelled diagrams to show what is meant by transverse waves and longitudinal waves.

Synoptic link

For more on information on electromagnetic waves, look at Topic P13.1.

Key words: oscillate, transverse wave, longitudinal wave, compression, rarefaction, mechanical waves, electromagnetic waves

On the up

To achieve the top grades, you should be able to describe evidence that, for ripples on a water surface and sound waves in air, it is the wave that travels, and not the water or the air itself.

- Waves transfer energy, not matter. We use waves to transfer energy and information. The direction in which the wave travels is the direction in which the wave transfers energy. Particles in the wave **oscillate** to and fro about a point; it is the whole wave that moves.
- There are different types of waves:
  - For a **transverse wave**, for example, a water wave, the direction of vibration of the particles is perpendicular to the direction in which the wave travels.

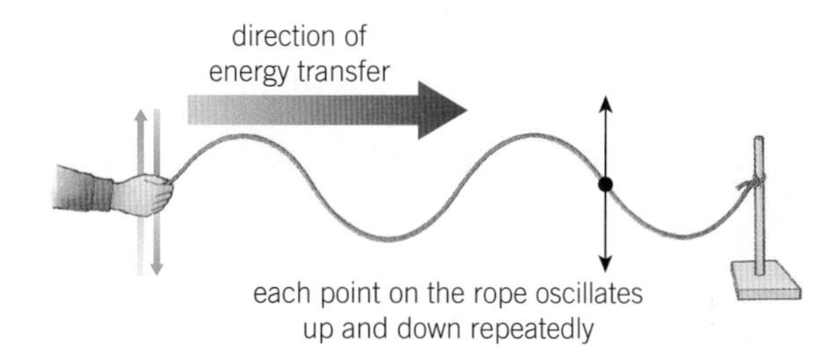

Transverse waves

  - For a **longitudinal wave**, for example, a sound wave, the direction of vibration of the particles is parallel to the direction in which the wave travels.
  - A longitudinal wave is made up of **compressions** and **rarefactions**.

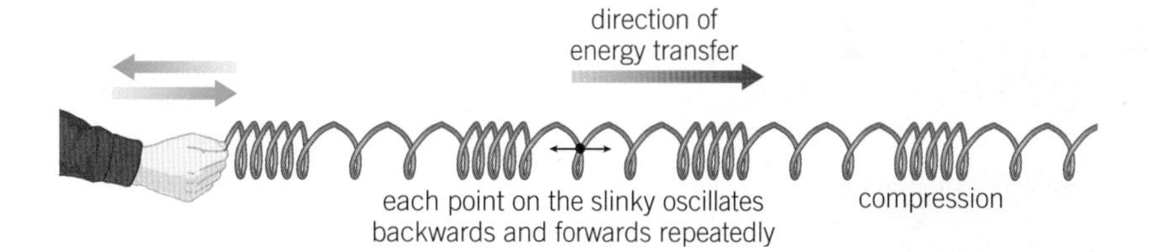

Longitudinal waves

- **Mechanical waves** travel through a medium (substance), for example, waves on springs and sound waves.
- **Electromagnetic waves** can travel through a vacuum, for example, light waves and radio waves.
- All electromagnetic waves are transverse waves.
- Mechanical waves may be transverse or longitudinal.

1. When a longitudinal wave passes through air, what happens to the air particles at a compression?
2. What type of wave can be produced on a stretched string?

Student Book **pages 176–177** P12

12.2 The properties of waves

Key points

- For any wave, its amplitude is the maximum displacement of a point on the wave from its undisturbed position, such as the height of the wave crest (or the depth of the wave trough) from the position at rest.
- For any wave, its frequency is the number of waves passing a point per second.
- The period of a wave $= \frac{1}{\text{frequency}}$
- For any wave, its wavelength is the distance from a point on the wave to the equivalent point on the next wave (e.g., from one wave trough to the next wave trough).
- The speed of a wave is $v = f \times \lambda$

Key words: amplitude, wavelength, frequency, speed

Study tip

Draw a labelled diagram to help you learn all of the terms associated with waves.

- The **amplitude** of a wave is the height of the wave crest, or the depth of the wave trough, from the midpoint of the wave.
- The greater the amplitude of a wave, the more energy it carries.
- The **wavelength** of a wave is the distance from one crest to the next crest or from one trough to the next trough.
- The **frequency** of a wave is the number of wave crests passing a point in one second. The unit of frequency is the hertz, Hz. This unit is equivalent to 'waves per second (/s)'.
- The period of a wave is the time it takes for one wavelength to pass a point.
- period of a wave $= \frac{1}{\text{frequency}}$
- The **speed** of a wave can be calculated using the equation $v = f\lambda$, where:

  v is the wave speed in metres per second, m/s

  f is the frequency in hertz, Hz

  λ is the wavelength in metres, m.
- The diagram below shows the wave features for a transverse wave, but all of the same terms apply to a longitudinal wave.
  - The wavelength of a longitudinal wave is the distance from the middle of one compression to the middle of the next compression (or from the middle of one rarefaction to the middle of the next rarefaction).
  - The frequency of a longitudinal wave is the number of compressions passing a point in one second.

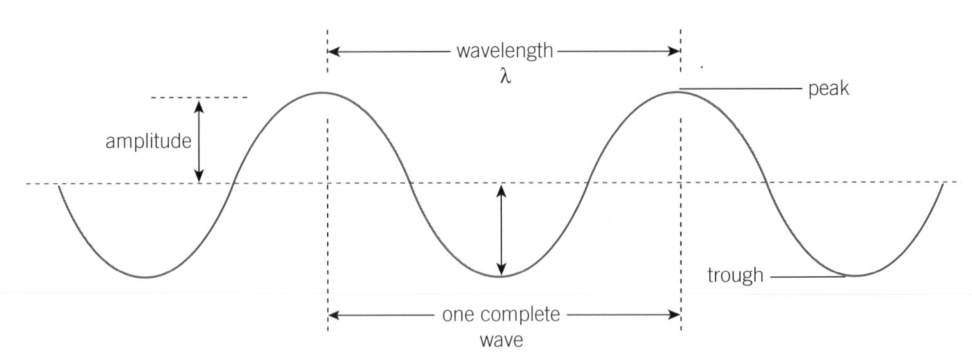

Wave properties

1 What is the speed of waves with a frequency of 5 Hz and a wavelength of 2 m?

2 What is the hertz equivalent to?

12.3 Reflection and refraction

- The behaviour of waves can be investigated with water waves in a ripple tank.
- Waves travelling towards a barrier or a boundary are called incident waves.
- If a barrier is placed in the tank, **reflection** takes place at the barrier. The reflected wavefront moves away from the barrier at the same angle to the barrier as the incident wavefront. There is no change in the speed or wavelength of the waves.

Wavefronts at a non-zero angle to the barrier

Incident wavefront

Reflected wavefront

Reflection of plane waves

- Waves change speed and wavelength when they cross a boundary between different substances. This can be seen in a ripple tank at the boundary between deep and shallow water. Unless the waves meet the boundary at right angles, the change in speed causes a change in direction. This effect is called **refraction**.

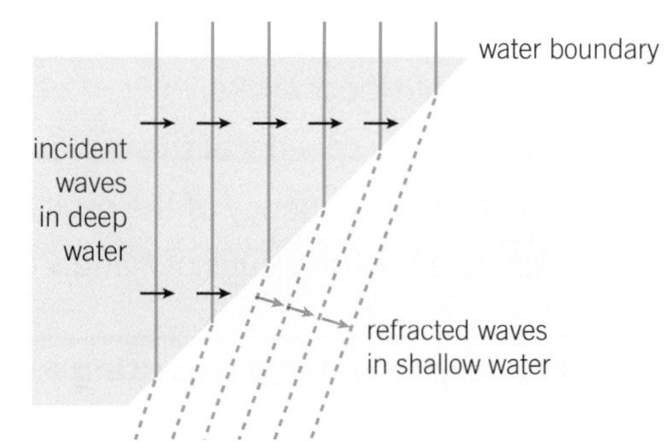

Refraction

- When waves meet a boundary with a different substance they may be:
  - totally or partially reflected
  - **transmitted** through the substance
  - absorbed by the substance.

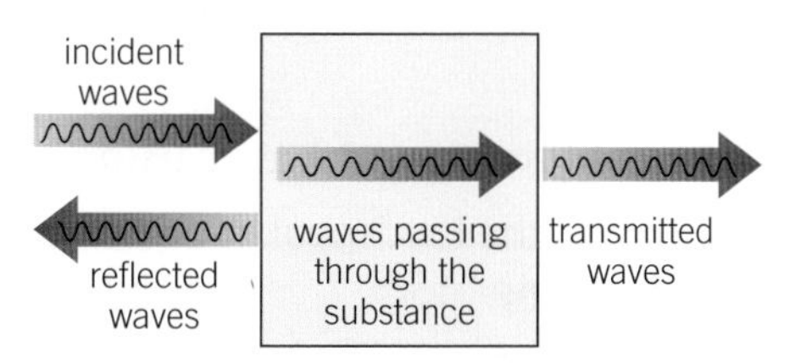

Waves and substances

1. Why does refraction take place?
2. What happens to the speed of waves when they are reflected?

Key points

- Plane waves in a ripple tank are reflected from a straight barrier at the same angle to the barrier as the incident waves because their speed and wavelength do not change on reflection.
- Plane waves crossing a boundary between two different materials are refracted unless they cross the boundary at normal incidence.
- Refraction occurs at a boundary between two different materials because the speed and wavelength of the waves change at the boundary.
- At a boundary between two different materials, waves can be transmitted or absorbed.

Synoptic link

For more information on the reflection and refraction of light, look at Topics P14.1 and P14.2.

Key words: reflection, refraction, transmitted

Study tip

Draw labelled diagrams to help you remember what happens when waves are reflected and refracted.

On the up

To achieve the highest grades, you should be able to explain refraction in terms of changes in the speed of waves as they cross a boundary between one medium and another.

Student Book pages 180–181 P12

12.4 More about waves

Key points

- Sound waves are vibrations that travel through a medium (a substance).
- Sound waves cannot travel through a vacuum (e.g., in outer space).
- To investigate waves, use:
  - a ripple tank for water waves
  - a stretched string for waves in a solid
  - a signal generator and loudspeaker for sound waves.

Key word: echo

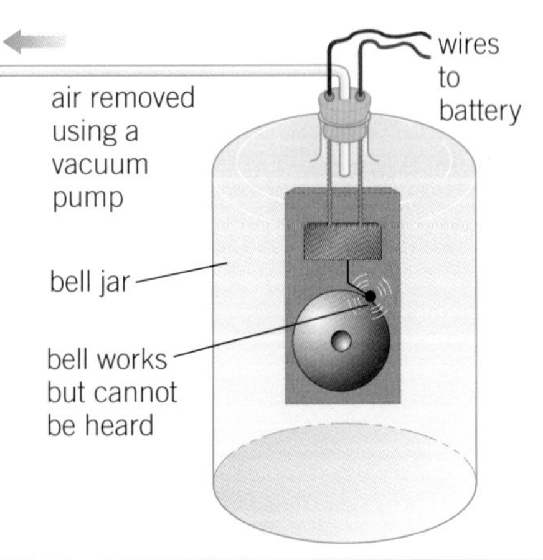

Sound waves can't travel through a vacuum

- Sound is caused by mechanical vibrations in a substance, and travels as a wave.
- Sound waves cannot travel through a vacuum (like space). This can be tested by listening to a ringing bell in a 'bell jar'. As the air is pumped out of the jar, the ringing sound fades away.
- Sound waves are longitudinal waves. The direction of the vibrations is parallel to the direction in which the wave travels.
- Sound waves can be reflected from hard, flat surfaces, such as walls, to produce an **echo**.

1 What is the reflection of a sound wave called?

2 Why can't sound waves travel through a vacuum?

Investigating waves

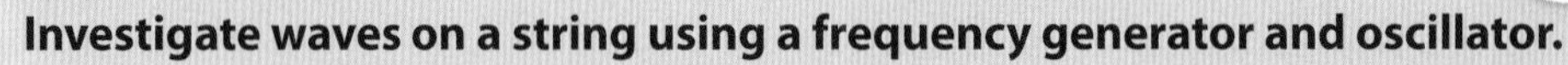

Investigate waves on a string using a frequency generator and oscillator.

Adjust the frequency of the oscillator until there is a single loop on the string, as shown opposite.

Note the frequency of the oscillator.

Measure the length L of the loop and calculate the wavelength λ of the wave. For a single loop, $\lambda = 2L$.

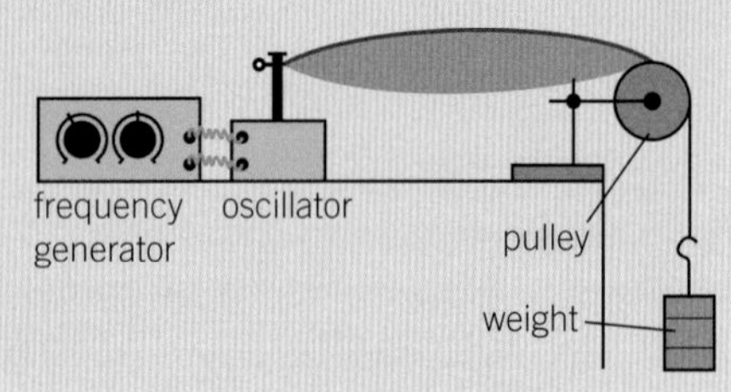

Investigating waves on a string

Calculate the speed v of the waves on the string using $v = f\lambda$.

Change the frequency of the oscillator to obtain more loops on the string.

Calculate further values of λ and v to see if the wave speed is the same.

Investigate water waves using a ripple tank.

Use a ruler to create plane waves.

Measure the time taken t for a wave to travel from one end of the tank to the other.

Measure the distance travelled s.

Calculate the speed of the water waves using $v = \frac{s}{t}$

Change the frequency of the waves by moving the ruler up and down faster.

Calculate further values of v to see if the wave speed is the same.

Investigate sound waves using a signal generator and a loudspeaker.

Observe the loudspeaker cone vibrating when it is connected to the signal generator.

Adjust the frequency of the oscillator and observe what happens to the movement of the loudspeaker cone.

Adjust the volume of the loudspeaker and observe what happens to the movement of the loudspeaker cone.

12.5 Sound waves

NOT THIS PAGE

Key points

- The pitch of a note increases if the frequency of the sound waves increases.
- The loudness of a note increases if the amplitude of the sound waves increases.
- Sound waves cause the ear drum to vibrate and the vibrations send signals to the brain.
- The conversion of sound waves to vibrations of solids only works over a limited range, so human hearing is limited.

Study tip

Draw diagrams to illustrate loud and quiet sounds with the same pitch and high and low sounds with the same loudness.

Key word: vibrate

- The pitch of a note depends on the frequency of the sound waves. The higher the frequency of the waves, the higher the pitch of the sound.
- The loudness of a sound depends on the amplitude of the sound waves. The greater the amplitude, the more energy the wave carries and the louder the sound.

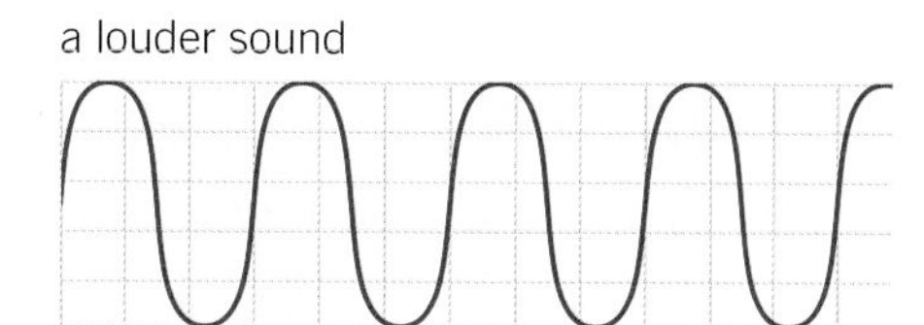

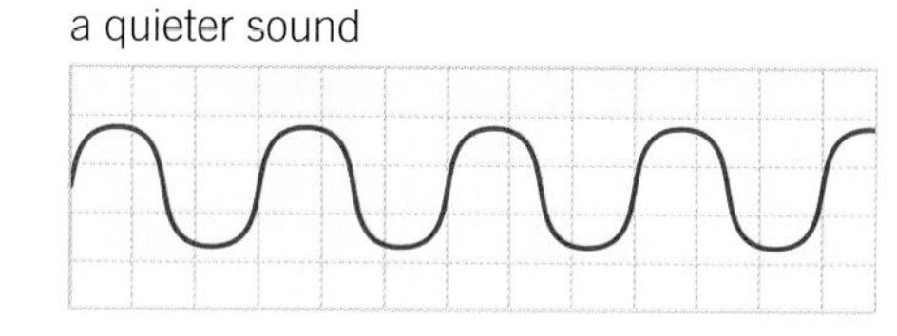

Sound waves

- Sound waves travel through the air. When they reach the ear, they make the eardrum **vibrate**. This sends an electrical signal to the brain.
- This conversion of the sound waves in the air to vibrations of the eardrum only works over a limited range of frequencies. The human ear can hear frequencies from about 20 Hz to about 20 kHz. The ability to hear the higher frequencies declines with age.
- The intensity of sounds that the human ear can detect varies with frequency. The normal ear hears best at a frequency of about 3 kHz, since the intensity of the least detectable sound is lowest at this frequency.

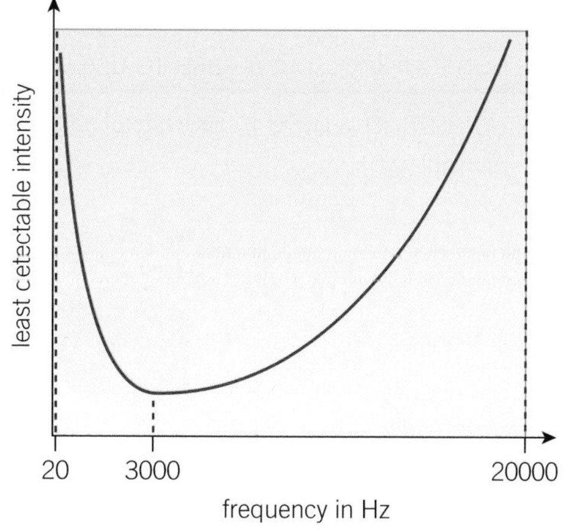

Frequency response of a normal ear

1. What happens to the pitch of a note as the frequency of the sound wave decreases?
2. What is the range of frequencies that can be heard by the human ear?

12.6 The uses of ultrasound

DONT NEED

Key points

- Ultrasound waves are sound waves with a frequency above 20 kHz.
- Ultrasound waves are partly reflected at a boundary between two different types of body tissue.
- Ultrasound waves reflected at boundaries are timed, and the timings are used to calculate distances.
- An ultrasound scan is non-ionising, so it is safer than an X-ray.

Key word: ultrasound wave

Study tip

Practise doing calculations on the distance travelled by ultrasound. Check carefully whether you are dealing with the distance travelled by the ultrasound or the distance to the boundary.

- The human ear can detect sound waves with frequencies between 20 Hz and 20 000 Hz. Sound waves of a higher frequency than this are called **ultrasound waves**.
- An ultrasound scanner uses an electronic device called a transducer to produce ultrasound waves. When a wave meets a boundary between two different materials, part of the wave is reflected. The wave travels back through the material to be detected by the transducer. The time it takes to reach the transducer can be used to calculate how far away the boundary is. The results are processed by a computer to give an image.
- The distance travelled by an ultrasound pulse can be calculated from the equation $s = v\,t$, where:

  s is the distance travelled in metres, m

  v is the speed of the ultrasound wave in metres per second, m/s

  t is the time taken in seconds, s.
- In the time taken between the transducer sending out a pulse of ultrasound and it returning, the ultrasound has travelled from the transducer to the boundary and back (that is, twice the distance to the boundary). So, the depth of the boundary below the surface is half the distance travelled by the ultrasound pulse.
- Ultrasound can be used in medicine for scanning. It is reflected at boundaries between different types of tissue, so it can be used for scanning soft tissue such as the eye and the kidney, and for scanning unborn babies. Ultrasound is non-ionising, so it is safer to use than X-rays.
- Ultrasound is used in industrial imaging (e.g., for detecting flaws in metal castings).

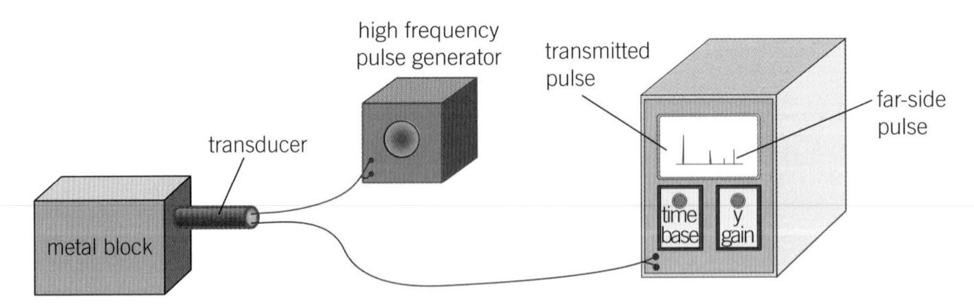

Detecting flaws in a metal

1. What is the minimum frequency of an ultrasound wave?
2. Why is an ultrasound scan safer than an X-ray?
3. An ultrasound pulse takes 3.9×10^{-5} s to travel a distance of 6.0 cm through a human kidney. Calculate the speed of ultrasound in the kidney.

a

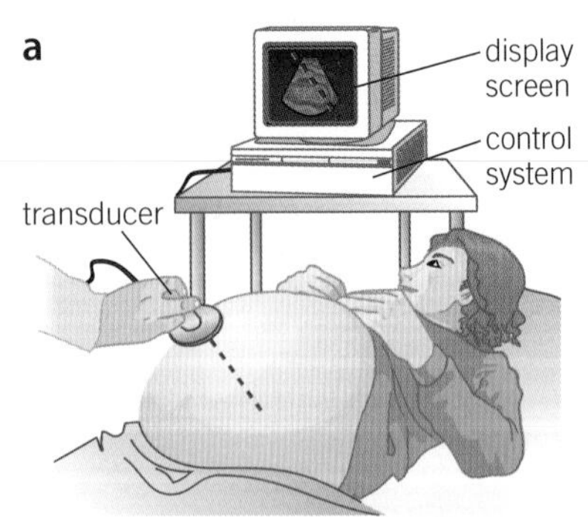

b

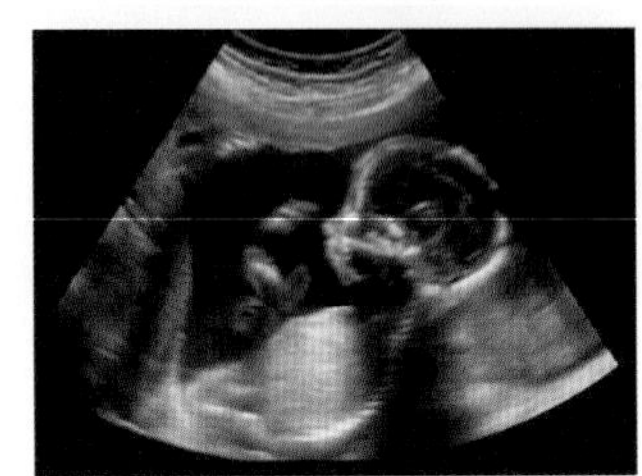

a *An ultrasound scanner system* **b** *An ultrasound image of a baby in the womb*

12.7 Seismic waves

DON'T NEED

Key points

- Seismic waves are waves that travel through the Earth.
- Seismic waves are produced in an earthquake and spread out from the epicentre.
- Primary seismic waves (P-waves) are longitudinal waves. Secondary seismic waves (S-waves) are transverse waves.
- The Earth has a solid inner core surrounded by a liquid outer core, which is surrounded by the Earth's mantle. The mantle is surrounded by the Earth's crust.

Synoptic links

For more information on transverse and longitudinal waves, look at Topic P12.1.

For more information on refraction, look at Topic P12.3.

Study tip

Draw labelled diagrams to show:

- the difference between P-waves and S-waves
- the structure of the Earth.

- Earthquakes are generated in the Earth's crust. The point an earthquake originates from is called the focus. The closest point on the Earth's surface to the focus is called the epicentre of the earthquake.
- An earthquake produces seismic waves that spread out from the epicentre and travel through the Earth.
- Primary seismic waves (P-waves) are longitudinal waves and cause the initial tremors. P-waves refract at the boundary between the mantle and the core.
- Secondary seismic waves (S-waves) are transverse waves. They travel more slowly than P-waves and cause tremors that are produced after the first minute or so. They cannot travel through the liquid core of the Earth.
- Both P-waves and S-waves bend as they travel though the mantle, because their speed gradually changes.
- Long waves (L-waves) travel more slowly than either P-waves or S-waves, and only travel in the Earth's crust. They arrive last and cause violent movement of the Earth's crust.

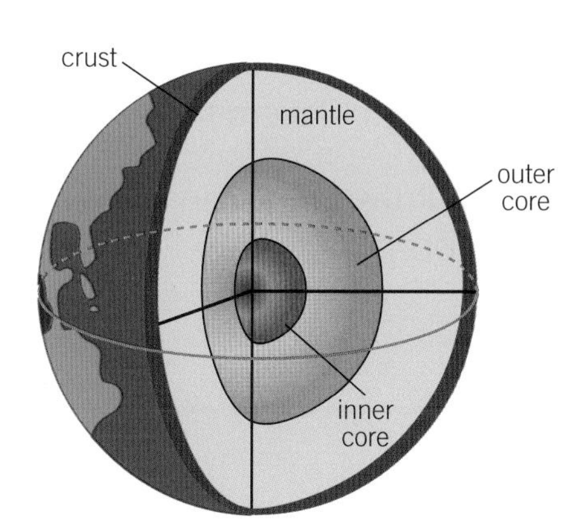

Inside the Earth

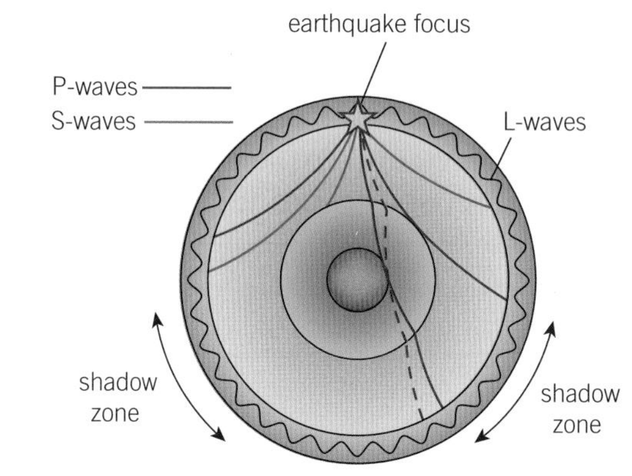

Refraction of seismic waves

- Earthquake activity is recorded by detectors on the surface called seismometers. Seismometer readings from different parts of the world are used to determine where the epicentre of an earthquake is.
- Analysis of seismic waves has allowed scientists to understand the structure of the Earth. The Earth has an inner core that is a liquid, surrounded by a solid outer core, which is surrounded by the Earth's mantle. The crust around the mantle is about 50 km deep.

1 Is the core of the Earth a solid, a liquid, or a gas?

2 What is the name given to the point an earthquake originates from?

Summary questions

1. Give one example of a mechanical wave. [1 mark]
2. Calculate the speed of waves with a frequency of 15 Hz and a wavelength of 0.45 m. [2 marks]
3. Define the frequency of a wave. [1 mark]
4. Describe how plane waves of different frequencies can be produced in a ripple tank. [2 marks]
5. Describe the difference between longitudinal waves and transverse waves. [2 marks]
6. Describe what is meant by the epicentre of an earthquake. [2 marks]
7. (H) Describe how you would demonstrate refraction using water waves in a ripple tank. [3 marks]
8. (H) What happens to the loudness of a sound as the amplitude of the sound wave decreases? [1 mark]
9. Explain what is meant by compression and rarefactions in a longitudinal wave. [2 marks]
10. (H) Give the frequency of sound to which the human ear is most sensitive. [1 mark]
11. What is the frequency of a wave of wavelength 2.5 cm travelling at 0.75 m/s? [4 marks]
12. (H) Give and explain two reasons why an ultrasound scan, rather than an X-ray, is used to produce an image of an unborn baby. [4 marks]

Chapter checklist

Tick when you have:

reviewed it after your lesson	✓		
revised once – some questions right	✓	✓	
revised twice – all questions right	✓	✓	✓

Move on to another topic when you have all three ticks

12.1 The nature of waves			
12.2 The properties of waves			
12.3 Reflection and refraction			
12.4 More about waves			
12.5 Sound waves			
12.6 The uses of ultrasound			
12.7 Seismic waves			

Student Book
pages 190–191

P13

13.1 The electromagnetic spectrum

Key points

- The electromagnetic spectrum (in order of decreasing wavelength and increasing frequency and energy) is made up of:
  - radio waves
  - microwaves
  - infrared radiation
  - visible light (red to violet)
  - ultraviolet waves
  - X-rays and gamma rays.
- The human eye can only detect visible light. The wavelength of visible light ranges from about 350 nm to about 650 nm.
- Electromagnetic waves transfer energy from a source to an absorber.
- The wave speed equation $v = f\lambda$ is used to calculate the frequency or wavelength of electromagnetic waves.

Synoptic link

For more information on properties of waves, look at Topics P12.1 and P12.2.

Key word: wave speed

Study tip

Use a mnemonic to help you remember the order of the waves in the electromagnetic spectrum. The sillier the mnemonic, the more likely you are to remember it!
For example:

Rude **M**edics **I**nject **V**errucas **U**sing **X**enon **G**as

- Electromagnetic waves are electric and magnetic disturbances. They travel as waves and transfer energy from a source to an absorber.
- All electromagnetic waves travel through space (a vacuum) at the same speed but they have different wavelengths and frequencies.
- All of the waves together are called the electromagnetic spectrum. We group the waves according to their wavelength and frequency.
- From longest wavelength (and smallest frequency) to shortest wavelength (and highest frequency), the electromagnetic spectrum consists of:
  - radio waves (with wavelengths up to 10^4 m)
  - microwaves
  - infrared
  - visible light
  - ultraviolet
  - X-rays
  - gamma rays (with wavelengths as short as 10^{-15} m)
- The spectrum is continuous. The frequencies and wavelengths at the boundaries are approximate as the different parts of the spectrum are not precisely defined.
- Different wavelengths of electromagnetic radiation are reflected, absorbed, or transmitted differently by different substances and types of surface.
- The higher the frequency of an electromagnetic wave, the more energy it transfers.
- All electromagnetic waves travel through space at a **wave speed** of 300 million m/s. You can calculate the wavelength or frequency of electromagnetic waves using the equation $v = f\lambda$, where:

  v is the wave speed (for all electromagnetic waves, $v = 3 \times 10^8$ m/s)

  f is the frequency in hertz, Hz

  λ is the wavelength in metres, m.

The wave speed equation can be rearranged to give $\lambda = \frac{v}{f}$ and $f = \frac{v}{\lambda}$

Because the speed of electromagnetic waves is so large, numbers for calculations are often expressed in standard form. For example, the speed of electromagnetic radiation in a vacuum is 3×10^8 m/s and the wavelength of a gamma ray could be 1.0×10^{-15} m.

1 Which part of the electromagnetic spectrum transfers the most energy?

2 Which part of the electromagnetic spectrum has the longest wavelength?

Student Book pages 192–193 P13

13.2 Light, infrared, microwaves, and radio waves

Key points

- White light contains all the colours of the visible spectrum.
- Infrared radiation is used for carrying signals from remote control handsets and inside optical fibres.
- Microwaves are used to carry satellite TV programmes and mobile phone calls.
- Radio waves are used for radio and TV broadcasting, radio communications, and mobile phone calls.
- Mobile phone radiation is microwave radiation and is also radio waves at near-microwave frequencies.
- Different types of electromagnetic radiation are hazardous in different ways. Microwaves and radio waves can heat the internal parts of people's bodies. Infrared radiation can cause skin burns.

Key word: white light

Study tip

Make a note of where and when you use light, infrared, microwaves, and radio waves in your home to help you remember some applications of these parts of the electromagnetic spectrum.

- Visible light is the part of the electromagnetic spectrum that is detected by our eyes. We see the different wavelengths within it as different colours. The wavelength increases across the spectrum as you go from violet to red. We see a mixture of all the colours as **white light**. Using the different shades and colours within visible light is a very important skill for photography.
- Infrared radiation (IR) is emitted by all objects. The hotter the object, the more infrared radiation it emits. Remote controls for devices such as TVs and DVD players use IR. IR can also be transmitted along optical fibres, and is used in special cameras that allow people to see in the dark.
- Microwaves are also used in communications. Microwave transmitters produce wavelengths that are able to pass through the atmosphere. They are used to send signals to and from satellites for TV programmes. Microwaves are also used in microwave ovens for heating food.
- Radio waves are used to transmit radio and TV programmes.
- Mobile phones use microwave radiation, and radio waves at near-microwave frequencies.
- Different parts of the electromagnetic spectrum are hazardous in different ways. Microwave radiation and radio waves penetrate your skin and are absorbed by body tissue, causing internal heating that may cause damage. Infrared radiation is absorbed by skin, and too much will cause you to get burnt.

Absorption and emission of infrared radiation

Take two identical cans, one with a light, shiny surface and one with a matt, black surface.

Fill each can with the same volume of hot water at the same temperature.

Leave the cans for the same length of time (20 minutes is about right), and measure the temperature in both cans.

The can with the lower temperature will have the surface that is the best emitter of infrared radiation.

Fill each can with the same volume of cold water at the same temperature.

Place both cans in the sunlight for the same length of time (30 minutes is about right), and measure the temperature of both cans.

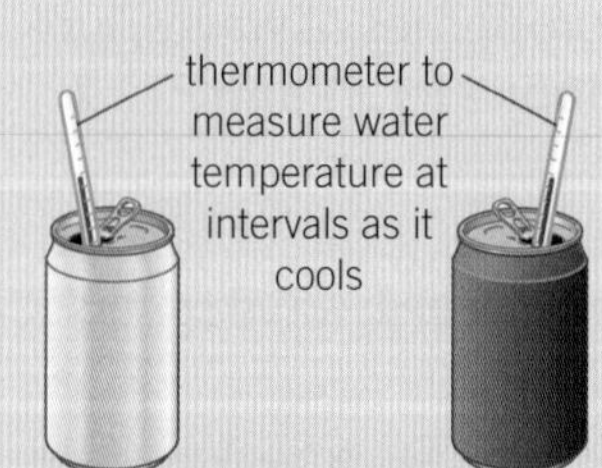

Testing different surfaces

The can with the higher temperature will have the surface that is the best absorber of infrared radiation.

1. Give one use of microwaves.
2. Which part of the electromagnetic spectrum is used to transmit TV programmes?

13.3 Communications

Key points

- Radio waves of different frequencies are used for different purposes because the wavelength (and so the frequency) of waves affects:
  - how far they travel
  - how much they spread
  - how much information they can carry.
- Microwaves are used for satellite TV signals.
- Further research is needed to evaluate whether or not mobile phones are safe to use.
- (H) Carrier waves are waves that are used to carry information. They do this by varying their amplitude.
- Optical fibres are very thin, transparent fibres that are used to transmit communication signals by light and infrared radiation.

Key word: carrier waves

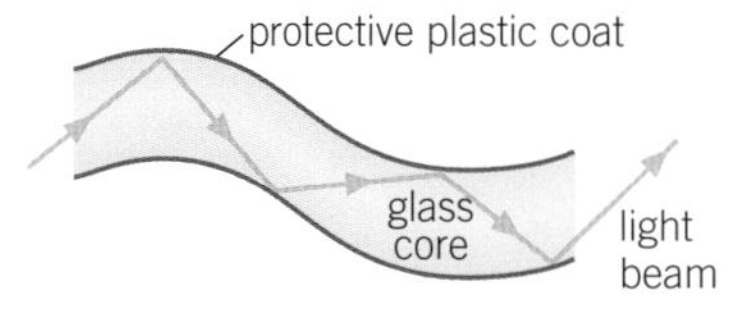

Waves in an optical fibre

Synoptic link

For more information on alternating current, look at Topic P5.1.

On the up

To achieve the top grades, you should be able to describe how carrier waves are used to transfer information.

- The waves used to carry any type of signal are called **carrier waves**.
- The radio and microwave spectrum is divided into different bands. The different bands are used for different communications purposes. This is because the shorter the wavelength of the waves:
  - the more information they carry
  - the shorter their range
  - the less they spread out.
- Microwaves are used for satellite phone and TV signals because they can travel between satellites in space and the Earth.
- Some scientists think that the radiation from mobile phones may affect the brain, especially in children, but further research is needed to evaluate this this claim.
- Optical fibres carrying visible light or infrared radiation are useful in communications because they carry much more information than radio wave and microwave transmissions (as light has a much shorter wavelength than radio waves or microwaves), and are more secure because the waves stay in the fibre.

1 Which types of waves are carried by optical fibres?

2 Which has the greater range, shorter wavelength radio waves or longer wavelength radio waves?

Higher

In a radio station, a microphone produces an audio signal when sound waves reach it.

The audio signal is used to modulate (vary the frequency of) carrier waves. These modulated carrier waves are supplied to the transmitter aerial, which emits radio waves that carry the audio signal.

When the radio waves are absorbed by a receiver aerial, they induce an alternating current, which causes oscillations in the receiver with the same frequency as the frequency of the radio waves.

The receiver circuit separates the audio signal from the carrier waves.

The audio signal is supplied to a loudspeaker, which emits sound waves.

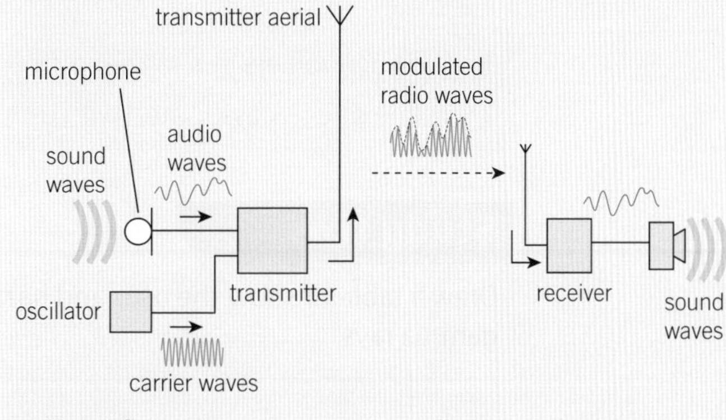

Using radio waves

Student Book pages 196–197 P13

13.4 Ultraviolet waves, X-rays, and gamma rays

Key points

- Ultraviolet waves have a shorter wavelength than visible light and can harm the skin and the eyes.
- X-rays are used in hospitals to make X-ray images.
- Gamma rays are used to kill harmful bacteria in food, to sterilise surgical equipment, and to kill cancer cells.
- Ionising radiation makes uncharged atoms become charged.
- X-rays and gamma rays damage living tissue when they pass through it.

Synoptic link

For more information on gamma rays, look at Topics P7.4 and P7.6.

Key word: ionisation

- Ultraviolet waves lie between visible light and X-rays in the electromagnetic spectrum. They can be used for security marking and in sunbeds. Ultraviolet waves can cause sunburn and skin cancer and may be harmful to human eyes.
- X-rays and gamma rays have similar frequencies, but they are produced in different ways.
- X-rays are used to make images of bones to check for fractures, and to detect cracks in metal objects.
- Gamma rays are used to kill harmful bacteria in food, which prevents food poisoning, and to sterilise surgical instruments. They may cause cancer but can also be used to kill cancer cells.
- When gamma rays and X-rays pass through substances, they knock electrons out of atoms in the substance, leaving the atoms with a positive charge. This process is called **ionisation**. Ionisation kills or damages living cells.
- Lead absorbs X-rays and gamma rays, so lead screens are often used to shield people who work with these types of radiation.
- People who work with ionising radiation wear a film badge. The radiation exposes the film in the badge so the worker's exposure can be monitored.

A film badge tells you how much ionising radiation the wearer has received

1 Why do workers in X-ray departments wear lead aprons?

2 Name one risk associated with using a sunbed.

Study tips

Draw a table to show the uses and hazards of ultraviolet waves, X-rays, and gamma rays.

13.5 X-rays in medicine

Key points

- X-rays are used in hospitals:
  - to make images of your internal body parts
  - to destroy tumours at or near the body surface.
- X-rays are ionising radiation and so can damage living tissue when they pass through it.
- X-rays are absorbed more by bones and teeth than by soft tissues.

Synoptic link

For more information on ionising radiation in medicine, look at Topic P7.6.

Key words: charge-coupled devices (CCDs), contrast medium, radiation dose

On the up

To achieve the top grades, you should be able to evaluate given data to draw conclusions about the risks and consequences of exposure to ionising radiation.

- X-rays can be used to diagnose some medical conditions. X-rays pass through soft tissue, but are absorbed by bones, teeth, and metal objects that are not too thin, so they are used to check for fractures and dental problems.
- X-ray photographs, formed on photographic film or flat **charge-coupled devices (CCDs)**, are used to examine your internal body parts.
- Some organs in the body that are made of soft tissue (e.g., the intestines) can be filled with a **contrast medium** that absorbs X-rays, so the organ can be seen on an X-ray image.

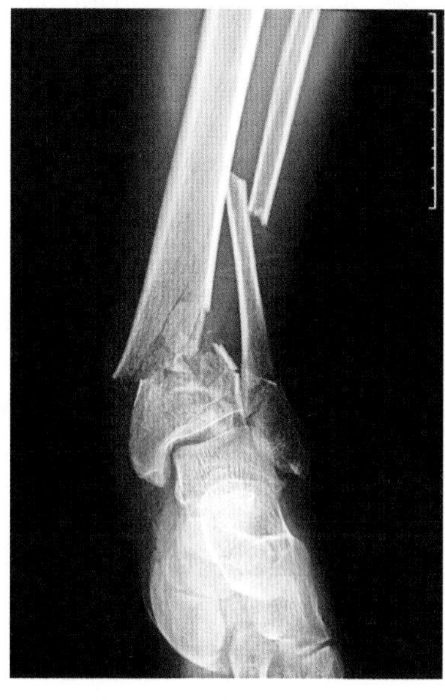

X-ray showing a broken bone

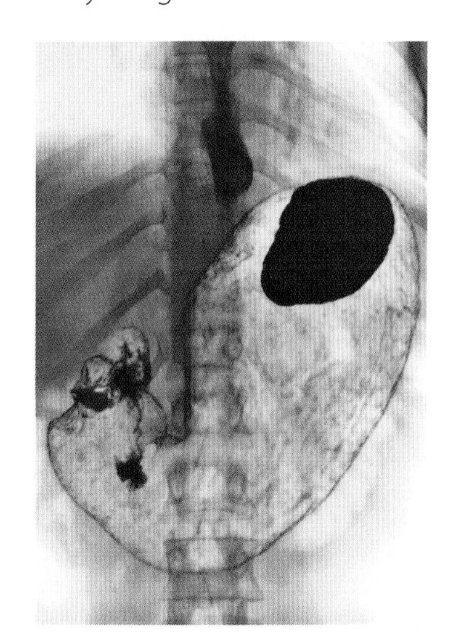

A coloured X-ray of a stomach ulcer

- X-rays cause ionisation and can damage living tissue when they pass through it. The **radiation dose** received is a measure of the damage done by the radiation.
- X-rays may also be used to destroy cancerous tumours near the surface of the body.

Low-energy X-rays are suitable for imaging, but they do not carry enough energy to destroy cancerous tumours, so high-energy X-rays are used for cancer treatment.

1 Why are X-rays used to diagnose broken bones?

2 What is radiation dose?

Summary questions

1. Identify which part of the electromagnetic spectrum transfers the least energy. [1 mark]
2. Describe what white light is. [1 mark]
3. Explain what an optical fibre is. [2 marks]
4. Describe what X-rays are used for in medicine. [3 marks]
5. Write the following parts of the electromagnetic spectrum in order of increasing wavelength:

   radio waves **visible light** **microwaves** [1 mark]
6. Explain how infrared cameras are used to take pictures in the dark. [3 marks]
7. Describe the hazards associated with ultraviolet light. [2 marks]
8. Explain two uses of gamma radiation. [4 marks]
9. Calculate the frequency of microwaves of wavelength 3 cm.
   Speed of electromagnetic radiation = 3×10^8 m/s. [3 marks]
10. Explain two advantages of using optical fibres in communications rather than radio waves or microwaves. [4 marks]
11. Explain why workers in hospital X-ray departments wear film badges. [3 marks]
12. Explain why a contrast medium is used to produce an X-ray image of a person's stomach. [3 marks]

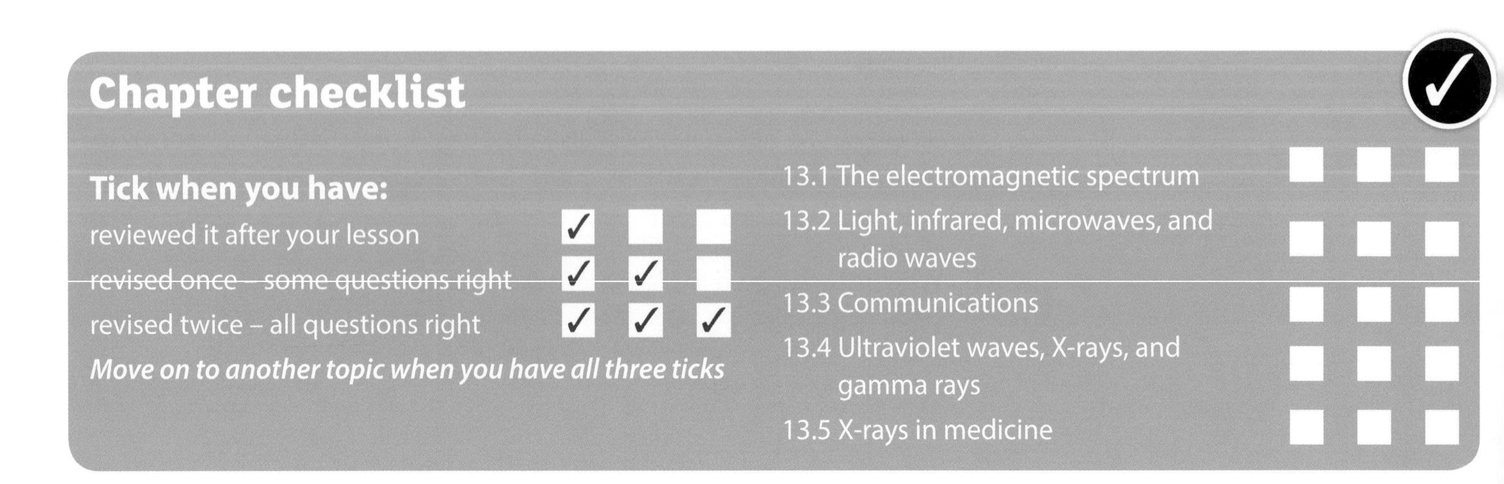

14.1 Reflection of light

Key points

- The normal at a point on a mirror is a line drawn perpendicular to the mirror.
- The law of reflection states that the angle of incidence = the angle of reflection.
- For a light ray reflected by a plane mirror:
  - the angle of incidence is the angle between the incident ray and the normal
  - the angle of reflection is the angle between the reflected ray and the normal.
- Specular reflection is reflection in a single direction without scattering. Diffuse reflection is reflection from a rough surface that scatters the light.

Synoptic link

For more information on reflection, look at Topic P12.3.

Study tip

Practise drawing diagrams to show how images are formed in a plane mirror.

- The image seen in a mirror is due to the reflection of light.
- The diagram shows how an image is formed by a plane (flat) mirror.

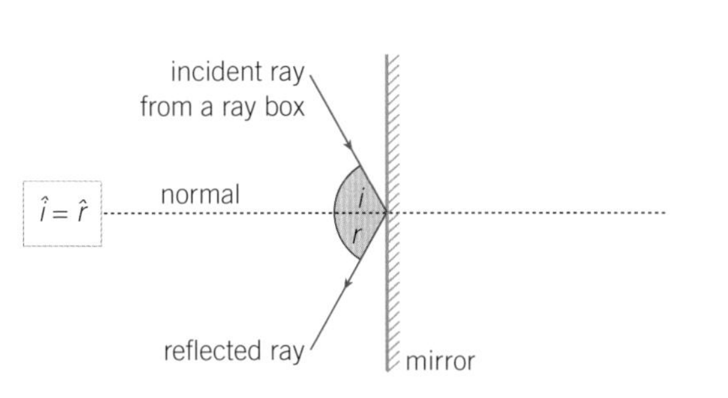

The law of reflection

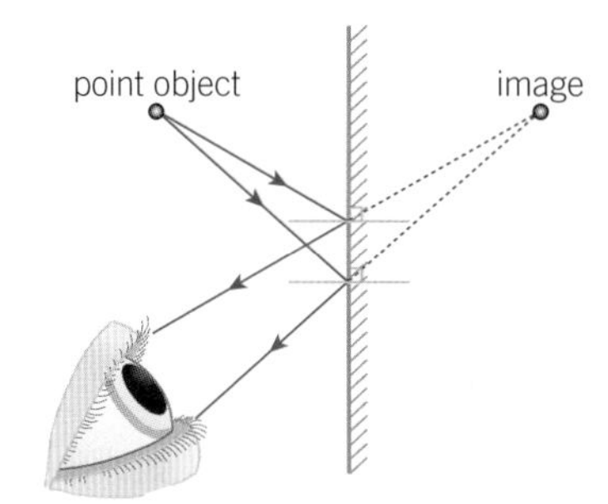

Image formation by a plane mirror

- We draw a line called the **normal** perpendicular to the mirror at the point where the incident ray hits the mirror.
- The **angle of incidence** is the angle between the incident ray and the normal.
- The **angle of reflection** is the angle between the reflected ray and the normal.
- For any reflected ray, the angle of incidence is equal to the angle of reflection.
- The image in a plane mirror is virtual, upright, the same size as the object, and the same distance behind the mirror as the object is in front.
- In a mirror, a **virtual image** is formed where rays of light appear to have come from. It cannot be formed on a screen.
- Reflection from a smooth surface, such as a mirror, is called **specular reflection** – parallel rays of light are reflected in a single direction.
- Reflection from a rough surface is called **diffuse reflection** – parallel rays of light are reflected in different directions.

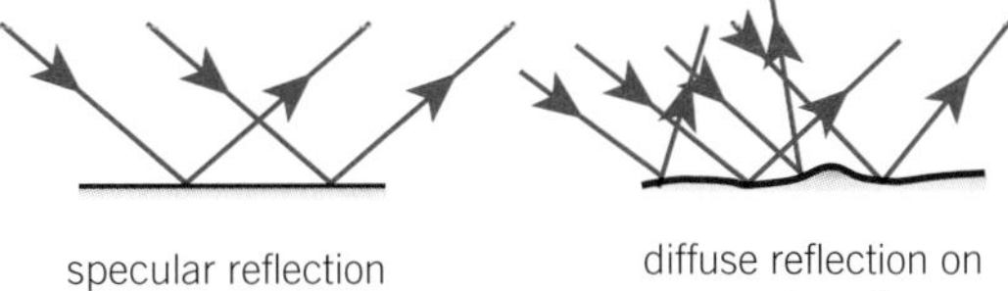

Reflection at a smooth and at a rough surface

1 What is the normal to a mirror?

2 What size is the image in a plane mirror?

Key words: normal, angle of incidence, angle of reflection, virtual image, specular reflection, diffuse reflection

14.2 Refraction of light

Key points

- Refraction is the change in direction of waves when they travel across a boundary from one medium to another.
- When a light ray refracts as it travels from air into glass, the angle of refraction is less than the angle of incidence.
- When a light ray refracts as it travels from glass into air, the angle of refraction is more than the angle of incidence.

Synoptic link

For more information on refraction, look at Topic P12.3.

Key word: refraction

Study tip

On ray diagrams, always draw in the normal.

On the up

To achieve the top grades, you should be able to draw diagrams of wave fronts changing speed as they travel from one medium to another, and use these to explain refraction.

1 Why does refraction take place?
2 Define the angle of refraction.

- **Refraction** is the change in direction of any kind of wave as it changes speed at a boundary.
- Since light is an electromagnetic wave, light is refracted as it passes between two transparent media.
- Light waves change speed when they cross a boundary, and hence are refracted. The change in speed causes a change in direction, unless the waves are travelling along a normal.
- The angle of refraction is the angle between the refracted ray and the normal.
- When a light ray travels from a less dense medium into a more dense medium, such as from air into glass, it is refracted towards the normal. The angle of refraction is less than the angle of incidence.
- When a light ray travels from a more dense medium into a less dense medium, such as from glass into air, it is refracted away from the normal. The angle of refraction is greater than the angle of incidence.

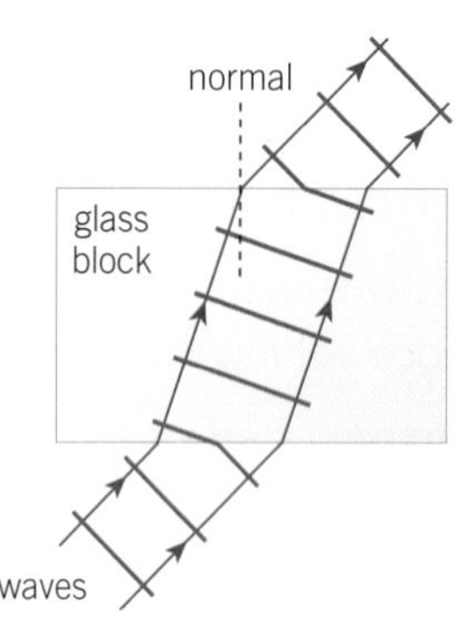

Refraction of waves

Investigating refraction of light

Set up a ray box and glass block as shown.

Draw around the glass block and mark the position of the incident ray and the refracted ray.

Remove the glass block and the ray box.

Draw the normal at the point where the incident ray enters the glass block.

Measure the angle of incidence and the angle of refraction.

normal
glass block
ray box

Refraction of light

Repeat with further angles of incidence and corresponding angles of refraction.

Compare the size of the angle of incidence with the size of the corresponding angle of refraction.

14.3 Light and colour

Key points

- The wavelength of light increases from violet to red across the visible spectrum.
- The colour of a surface depends on the pigments of the surface materials and the wavelength of the light the pigments absorb.
- A translucent object lets light pass through it but scatters or refracts the light inside it.
- A transparent object lets all the light that enters it pass through it and does not scatter or refract the light inside the object.

Synoptic link

For more information on the electromagnetic spectrum, look at Topic P13.1.

Study tip

Use coloured pencils to practise drawing diagrams to show the absorption and reflection of light by surfaces of different colours.

- The visible light spectrum, which can be detected by the human eye, is a very narrow part of the electromagnetic spectrum. The wavelengths in the visible spectrum increase from violet to red.
- The human eye sees different frequencies of light as different colours. The frequencies form a continuous spectrum, so one colour merges into another.
- Colour filters absorb certain wavelengths and transmit others.
- Pigments in materials absorb light of specific wavelengths and strongly reflect other wavelengths.
- A white surface has no pigments, so it reflects light of any wavelength either partially or totally. It reflects all the colours that make up white light. A black surface absorbs all wavelengths and reflects none.

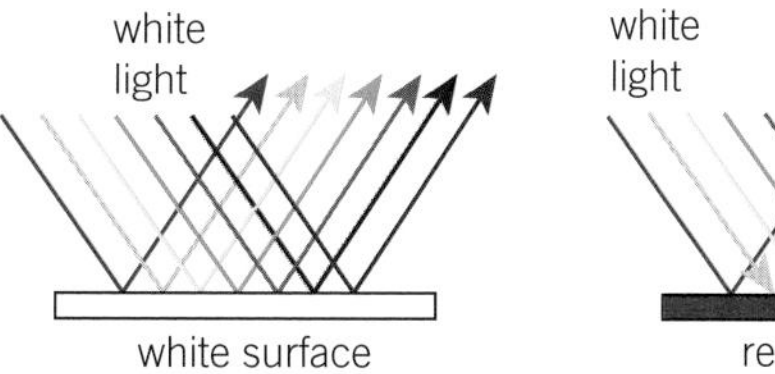

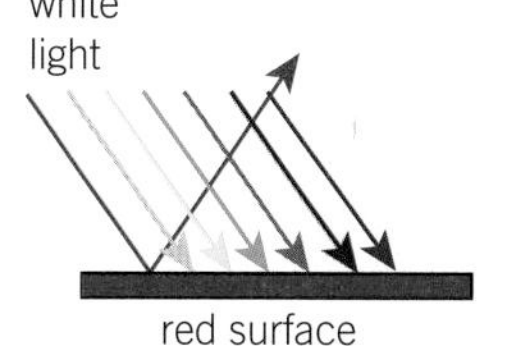

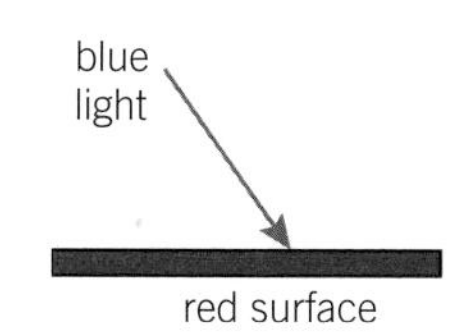

Surfaces and colour

- Transparent objects transmit all the light incident on them. No light is absorbed at the surface.
- Translucent objects allow light to pass through, but it is scattered or refracted inside them.
- Opaque objects absorb all the light incident on them.

Surface tests

Set up a ray box on a blank sheet of paper.

Shine a ray of light along the paper onto a plane mirror.

Mark the positions of the mirror, the incident ray, and the reflected ray.

Remove the mirror and draw in the normal at the point where the incident ray meets the mirror.

Remove the ray box and mirror and measure the angle of incidence and the angle of reflection.

Repeat for other angles of incidence and corresponding angles of reflection.

Replace the mirror with different coloured surfaces and repeat.

1 What is meant by the spectrum of visible light?

2 What is meant by a transparent object?

14.4 Lenses

Key points

- A convex lens focuses parallel rays to a point called the principal focus.
- A concave lens makes parallel rays spread out as if they had come from a point called the principal focus.
- A real image is formed by a convex lens if the object is further away than the principal focus. A virtual image is formed by a convex lens if the object is nearer than the principal focus.
- magnification = $\frac{\text{image height}}{\text{object height}}$

Key words: convex lens, principal focus, focal length, real image, virtual image, magnifying glass, magnification, concave lens

Converging lens

- Parallel rays of light that pass through a **convex lens** (converging lens) are refracted so that they converge to a point. This point is called the **principal focus** (focal point). The distance from the centre of the lens to the principal focus is the **focal length**.
- Because light can pass through the lens in either direction, there is a principal focus on either side of the lens.
- If the object is further away from the lens than the principal focus, an inverted, **real image** is formed that is smaller than the object.
- If the object is nearer to the lens than the principal focus, an upright, **virtual image** is formed behind the object. The image is now magnified – the lens acts as a **magnifying glass**.
- The **magnification** can be calculated using

$$\text{magnification} = \frac{\text{image height}}{\text{object height}}$$

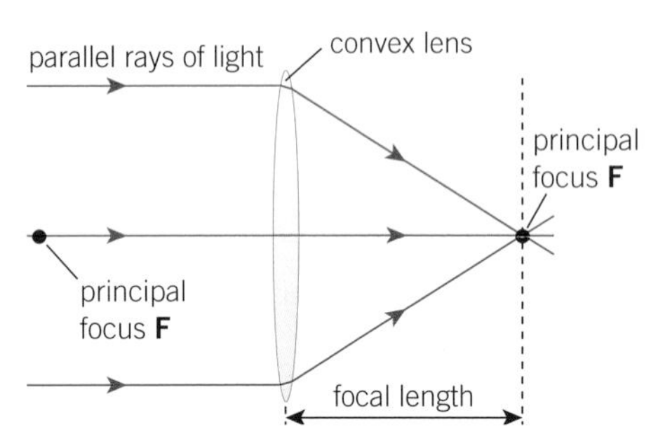

The focal length of a convex lens

Diverging lens

- Parallel rays of light that pass through a **concave lens** (diverging lens) are refracted so that they diverge away from a point. This point is called the principal focus.
- The distance from the centre of the lens to the principal focus is the focal length.
- Because light can pass through the lens in either direction, there is a principal focus on either side of the lens.
- The image produced by a diverging lens is always virtual.

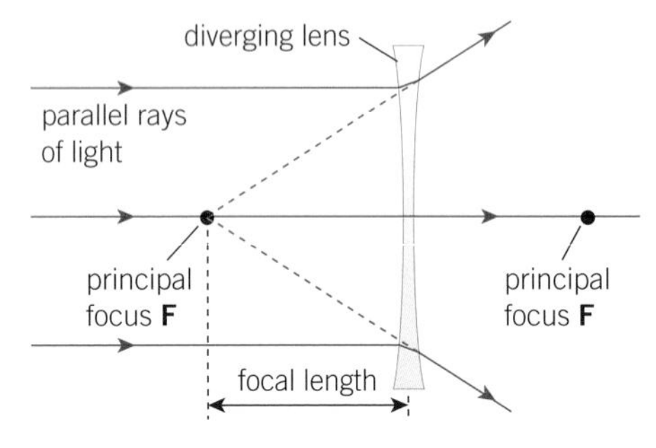

The focal length of a concave lens

1. What is the principal focus of a converging (convex) lens?
2. How can a converging lens be made to produce a virtual image?

Student Book pages 210–211 P14

14.5 Using lenses

Key points

- A ray diagram can be drawn to find the position and nature of an image formed by a lens.
- When an object is placed between a convex lens and its principal focus *F*, the image formed is virtual, upright, magnified, and on the same side of the lens as the object.
- A camera contains a convex lens that is used to form a real image of an object.
- A magnifying glass is a convex lens that is used to form a virtual image of an object.

Study tip

Practise drawing ray diagrams. Use a ruler and make your diagrams neat.

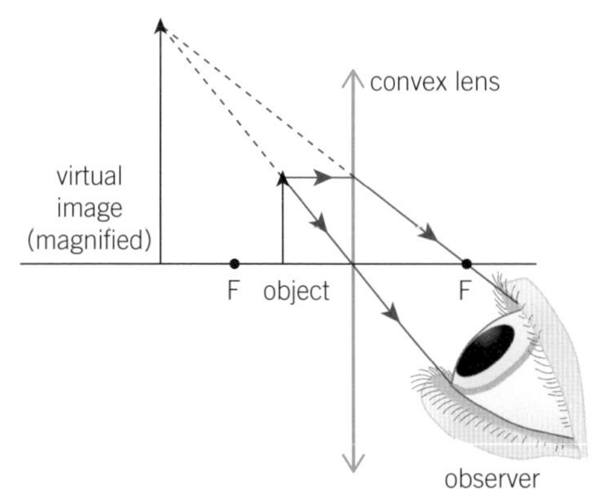

Formation of a virtual image by a convex lens

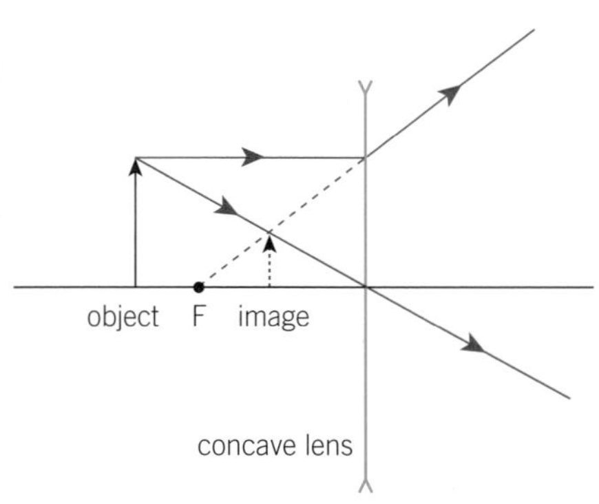

Image formation by a concave lens

- You can draw ray diagrams to find the image that different lenses produce with objects in different positions.
- The line through the centre of the lens and at right angles to it is called the principal axis. This should be included in the diagram.
- Ray diagrams use three construction rays from a single point on the object to locate the corresponding point on the image. For a convex lens:
  - a ray parallel to the principal axis is refracted through the principal focus
  - a ray through the centre of the lens travels straight on
  - a ray through the principal focus is refracted parallel to the principal axis.

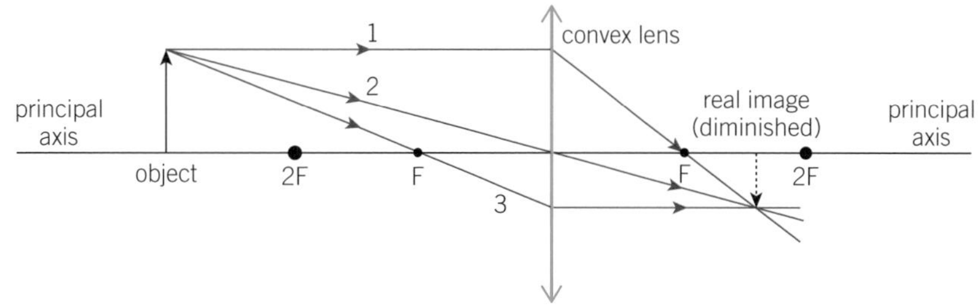

ray 1 is parallel to the axis and is refracted through F
ray 2 passes straight through the centre of the lens
ray 3 passes through F and is refracted parallel to the axis

Formation of a real image by a convex lens

- A camera uses a converging lens to form a real image of an object on a film or an array of pixels.
- When an object is placed closer to the convex lens than the principal focus, the lens forms a virtual image of the object that is upright, magnified, and on the same side of the lens as the object.
- A magnifying glass uses a convex lens to produce a virtual, magnified image in this way.

1 What are construction rays?
2 Is the image formed in a camera real or virtual?

On the up

To achieve the top grades, you should be able to construct ray diagrams to determine the position and nature of the image produced by concave and convex lenses, with the object at different distances from the lens.

Summary questions

1. Suggest what type of image is formed in a plane mirror. [1 mark]
2. Describe the angle of incidence. [1 mark]
3. Identify what happens to the speed of a light wave as it travels from air into a glass block. [1 mark]
4. Name what sort of lens is used in a magnifying glass. [1 mark]
5. Explain the difference between a real image and a virtual image. [2 marks]
6. Explain what happens to a light wave as it travels from air into a glass block. [2 marks]
7. Explain what is meant by the normal. [2 marks]
8. Describe an opaque surface. [2 marks]
9. Explain why a white surface appears white. [3 marks]
10. Describe what is meant by the focal length of a lens. [2 marks]
11. When a lens is used as a magnifying glass, the image produced is 8.4 cm high. If the magnification is 2.4, calculate the height of the object. [2 marks]
12. Sketch a diagram to show how three construction rays can be used to determine the position of the image in a converging lens. [3 marks]

Chapter checklist

Tick when you have:

reviewed it after your lesson	✓		
revised once – some questions right	✓	✓	
revised twice – all questions right	✓	✓	✓

Move on to another topic when you have all three ticks

14.1 Reflection of light			
14.2 Refraction of light			
14.3 Light and colour			
14.4 Lenses			
14.5 Using lenses			

15.1 Magnetic fields

Key points

- Like poles repel and unlike poles attract.
- The magnetic field lines of a bar magnet curve around from the north pole of the bar magnet to the south pole.
- Induced magnetism is magnetism created in an unmagnetised magnetic material when the material is placed in a magnetic field.
- Steel is used instead of iron to make permanent magnets because steel does not lose its magnetism easily but iron does.

Study tip

Draw diagrams to show the magnetic field around a bar magnet. Make sure you add arrows to the field lines pointing away from the north pole and towards the south pole.

Key word: magnetic field

- Any iron or steel object can be magnetised (or demagnetised if it's already magnetised). A few other materials, such as cobalt and nickel, can also be magnetised and demagnetised.
- Permanent magnets are usually made of steel because magnetised steel does not lose its magnetism easily.
- The region around a magnet, in which a magnetic material experiences a force, is called a **magnetic field**.
- The magnetic field lines around a bar magnet are drawn coming from the north pole of the magnet and curving around into the south pole.
- When two north poles are placed together, or two south poles are placed together, they will repel each other.

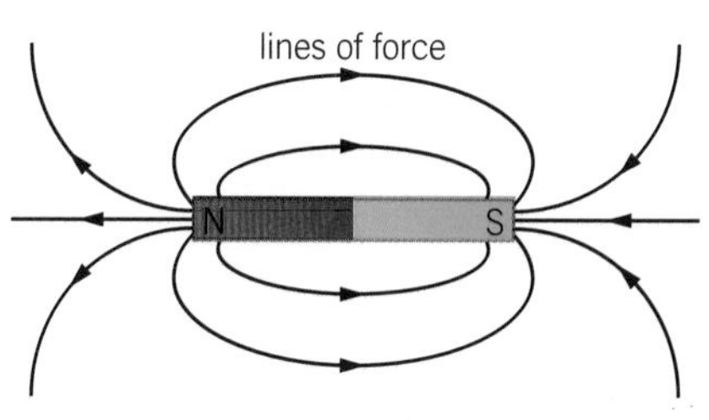

The magnetic field of a bar magnet

- When a north pole and a south pole are placed together they will attract each other.
- If an unmagnetised magnetic material is placed in a magnetic field it will become magnetised. This is called induced magnetism.
- Induced magnetism will always cause a force of attraction between the existing magnet and the unmagnetised magnetic material placed in the magnetic field.

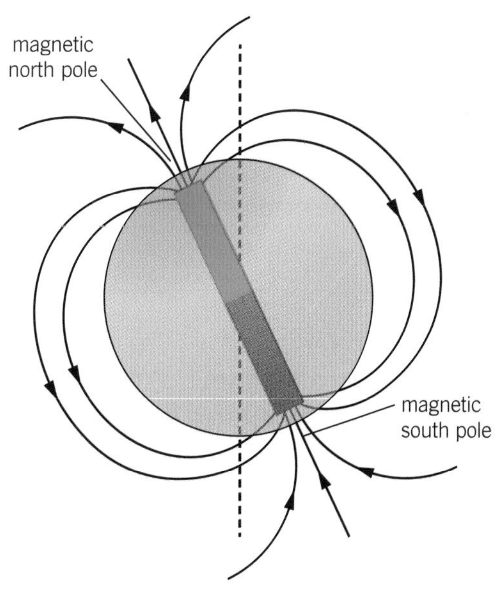

The Earth's magnetic field. Scientists have plotted the Earth's magnetic field accurately. The pattern is like that of a bar magnet. But the Earth is partly molten inside, and this helps to explain why the magnetic poles move about gradually

1 Give two examples of magnetic materials.
2 What material is used to make permanent magnets?

Student Book pages 216–217 P15

15.2 Magnetic fields of electric currents

Key points

- The magnetic field lines *around a wire* are circles, centred on the wire, in a plane perpendicular to the wire.
- The magnetic field lines *in a solenoid* are parallel to its axis and are all in the same direction. A uniform magnetic field is one in which the magnetic field lines are parallel.
- Increasing the current makes the magnetic field stronger. Reversing the direction of the current reverses the magnetic field lines.
- An electromagnet is a solenoid that has an iron core. It consists of an insulated wire wrapped around an iron bar.

Synoptic link

For more information on electric current, look at Topic P4.2.

Key words: solenoid, electromagnet

- As an electric current passes along a wire, a magnetic field is induced around the wire.
  - The magnetic field lines around the wire are a series of concentric circles that are centred on the wire and are perpendicular to the wire.
  - The corkscrew rule gives the direction of the magnetic field for each direction of the current.

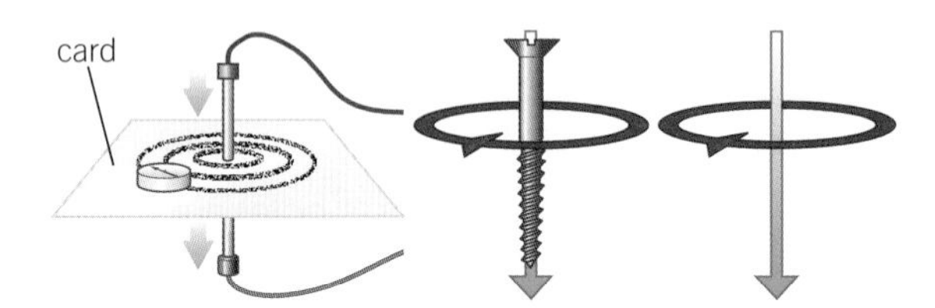

The corkscrew rule

- A **solenoid** is a long coil of insulated wire.
  - The magnetic field lines inside the solenoid are parallel to the axis of the solenoid and are all in the same direction (that is, uniform).
  - The magnetic field around a solenoid is stronger than the field around a straight wire.
  - The magnetic field outside the solenoid is like the field of a bar magnet, but the lines are complete loops as they pass through the centre of the solenoid.
- For a single wire, and a solenoid:
  - Reversing the direction of current reverses the direction of the magnetic field.
  - Increasing the current increases the strength of the magnetic field.
- An **electromagnet** is a solenoid in which an insulated wire is wrapped around an iron bar. Switching the current through the solenoid off causes the iron bar to lose its magnetism.

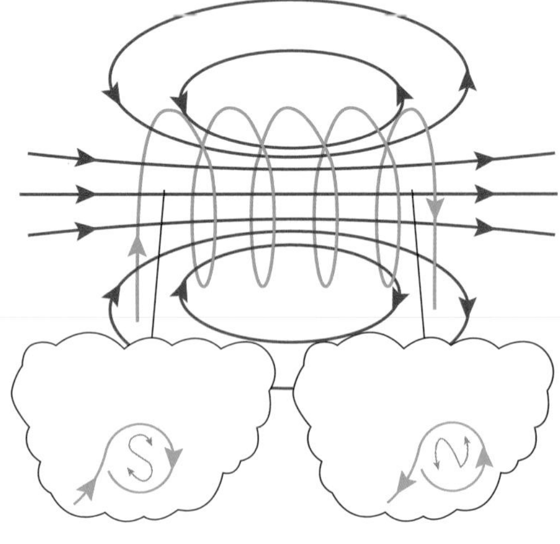

The magnetic field of a solenoid

1 What shape is the magnetic field around a current-carrying wire?
2 What is a solenoid?

Student Book pages 218–219

P15

15.3 Electromagnets in devices

Key points

- Electromagnets are used in scrapyard cranes, circuit breakers, electric bells, and relays.
- An electromagnet works in a circuit breaker or electric bell or a relay by attracting an iron armature, which opens a switch.

- Electromagnets are useful because they allow a magnetic field to be switched on and off by switching the current on and off.
- Electromagnets are used in many devices, including scrapyard cranes, circuit breakers, electric bells, and relays.
- In a scrapyard crane, a current is switched on to attract the steel body of a vehicle to the electromagnet in the crane. The crane moves the vehicle, and then current can then be switched off to release the vehicle.
- In a circuit breaker, too great a current makes the electromagnet open a switch, which cuts the current off. The switch then needs to be reset manually.

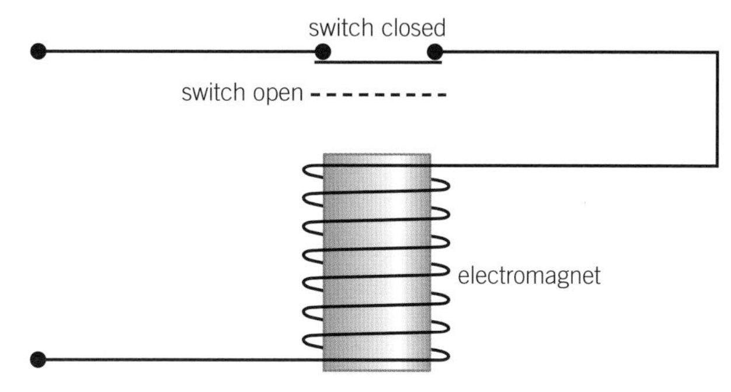

A circuit breaker

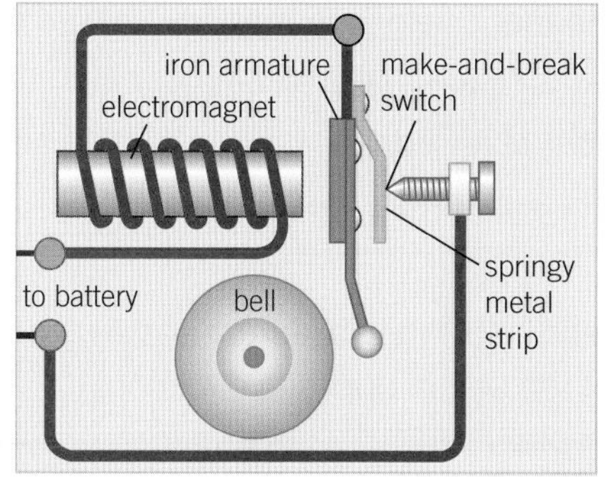

An electric bell

- When an electric bell is pushed on, an electromagnet attracts an iron armature, which hits a bell. Because the iron armature is pulled away from the make-and-break switch, the electromagnet loses it magnetism and the armature springs back. This process repeats itself rapidly.
- A relay uses the small current through the coil of an electromagnet to attract an iron armature, which switches on a much larger current in another circuit.

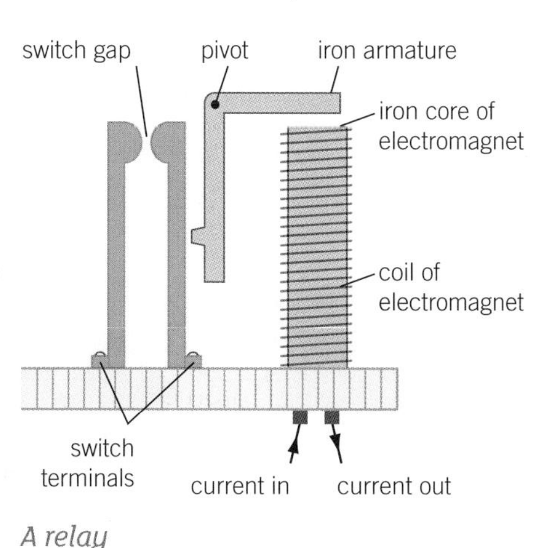

A relay

On the up

To achieve the top grades, you should be able to interpret diagrams of electromagnetic devices to explain how they work.

1 What is an electromagnet?

2 What is a relay?

15.4 The motor effect

Key points

- In the motor effect, the force is:
  - increased if the current or the strength of the magnetic field or the length of the conductor is increased
  - reversed if the direction of the current or the magnetic field is reversed.
- An electric motor has a coil that turns when a current is passed through it.
- Magnetic flux density is a measure of the strength of a magnetic field.
- To calculate the force on a current-carrying conductor at right angles to the lines of a magnetic field, use the equation $F = B I l$

Synoptic link

For more information on forces, look at Topic P8.3.

Key words: motor effect, Fleming's left-hand rule

Study tip

Draw a set of diagrams to show the force on the coil in a motor during different parts of the rotation of the coil.

- A wire carrying an electric current inside a magnetic field may experience a force. This is called the **motor effect**. The force is at its maximum if the wire is at an angle of 90° to the magnetic field. The force is zero if the wire is parallel to the magnetic field. **Fleming's left-hand rule** is used to determine the direction of the force.

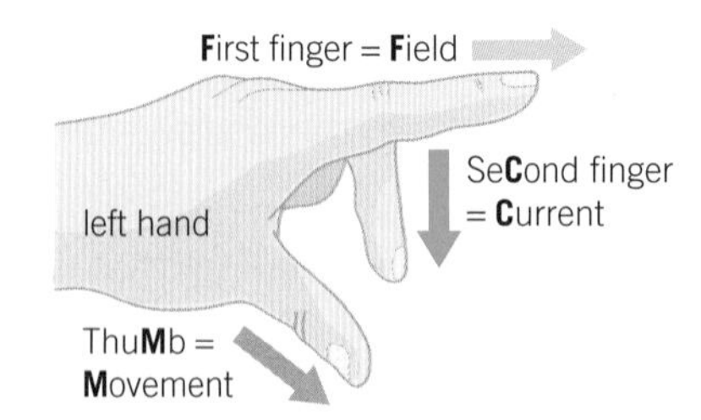

Fleming's left-hand rule

- The magnetic flux density of a magnetic field is a measure of how strong the magnetic field is. Its symbol is B, and its unit is the tesla (T).
- The size of the force on a current-carrying conductor in a magnetic field can be calculated using the equation $F = B I l$, where:

  F is the force on the conductor in newtons, N
  B is the magnetic flux density in teslas, T
  I is the current in amperes, A
  l is the length of the conductor in the magnetic field in metres, m.
- The direction of the force on the wire is reversed if either the direction of the current or the direction of the magnetic field is reversed.
- The force on the wire is increased if either the size of the current or the strength of the magnetic field is increased.
- An electric motor contains a coil in a magnetic field. When a current passes through the coil, a force acts on each side of the coil, in opposite directions due to the motor effect. This makes the coil spin.
- The split-ring commutator reverses the direction of the current around the coil every half turn so that the coil is always pushed in the same direction.

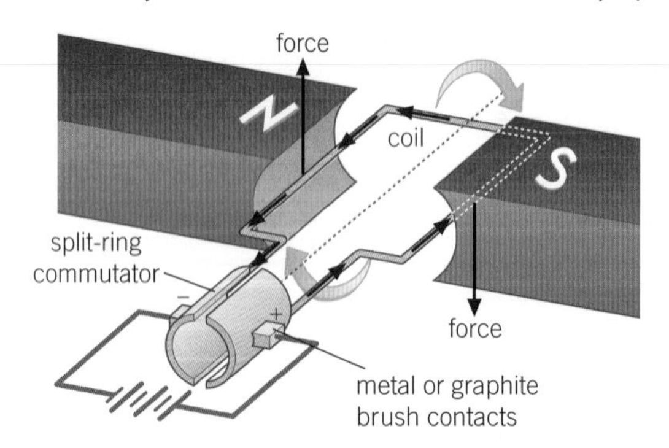

The electric motor

1 What happens to the direction of the force on a current-carrying wire if the direction of the current and the magnetic field are both reversed?

2 What is the unit of magnetic flux density?

15.5 The generator effect

Key points

- The generator effect is the effect of inducing a potential difference using a magnetic field.
- When a conductor crosses through the lines of a magnetic field, a potential difference is induced across the ends of the conductor.
- The faster a conductor crosses through the lines of a magnetic field, the bigger is the induced potential difference. When a direct-current electromagnet is used, it needs to be switched on or off to induce a potential difference.
- The direction of an induced current always opposes the original change that caused it.

Synoptic link

For more information on potential difference, look at Topic P4.3.

Study tip

Make a table to show the factors that increase the size of the induced p.d. and that decrease the size of the induced p.d., and the factors that affect the direction of the induced p.d.

Key words: electromagnetic induction, generator effect

- If an electrical conductor 'cuts' through magnetic field lines, a potential difference (p.d.) is induced across the ends of the conductor.
- If a magnet is moved into a coil of wire, a p.d. is induced across the ends of the coil. This process is called **electromagnetic induction**. If the wire or coil is part of a complete circuit, a current passes through it. This effect is called the **generator effect**.

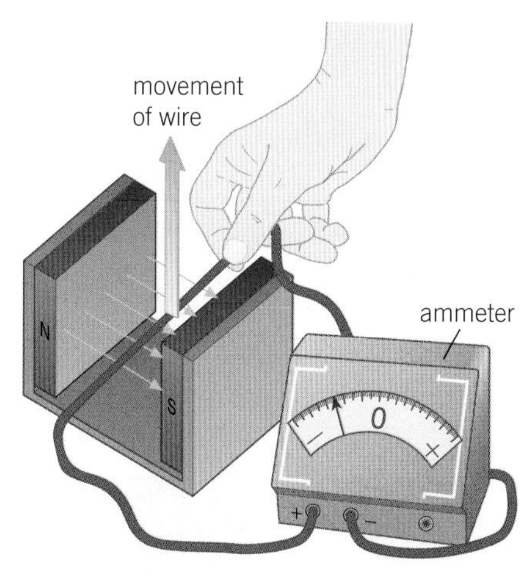

The generator effect

- If the direction of movement of the wire or coil is reversed, or the polarity of the magnet is reversed, the direction of the induced p.d. is also reversed. A p.d. is only induced while there is a changing magnetic field (that is, while the magnet and the coil are moving in relation to one another).

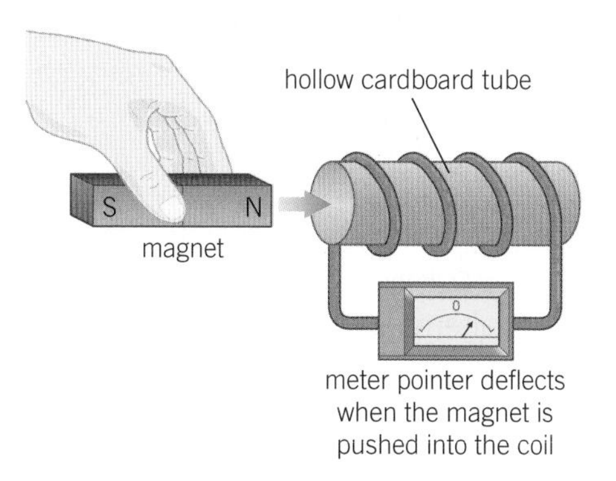

Testing the generator effect

- The size of the induced p.d. is increased by increasing:
  - the speed of movement
  - the strength of the magnetic field
  - the number of turns on the coil.

1 Why is there no potential difference induced when a bar magnet is held stationary inside a coil of wire?

2 What is the effect on the induced p.d. of reversing the direction of the current in a conductor cutting magnetic field lines?

Student Book pages 224–225 P15

15.6 The alternating-current generator

Key points

- A simple a.c. generator is made up of a coil that spins in a uniform magnetic field.
- The waveform, displayed on an oscilloscope, of the a.c. generator's induced potential difference is at:
  - its peak value when the sides of the coil cross directly through the magnetic field lines
  - its zero value when the sides of the coil move parallel to the field lines.
- A simple d.c. generator has a split-ring commutator instead of two slip rings.

Synoptic link

For more information on alternating potential difference, look at Topic P5.1.

Key words: alternator, dynamo

Study tip

Draw p.d.–time graphs to show what happens when an alternator and a dynamo are rotated at different speeds. Remember that the faster the coil rotates inside the magnetic field, the higher the peak value of the p.d. induced across the coil and the higher the frequency of the alternating current induced in the coil.

- A simple **alternator** is an alternating-current generator. It consists of a rectangular coil that spins in a uniform magnetic field. The potential difference (p.d.) generated is alternately positive and negative.
- The size of the induced p.d. is greatest when the plane of the coil is parallel to the direction of the magnetic field, as the coil is then 'cutting' the magnetic field lines.
- The size of the induced p.d. is zero when the plane of the coil is perpendicular to the magnetic field lines, as the coil is then moving in the same direction as the magnetic field lines.
- The faster the coil rotates, the higher the frequency and the greater the peak value of the alternating current produced.

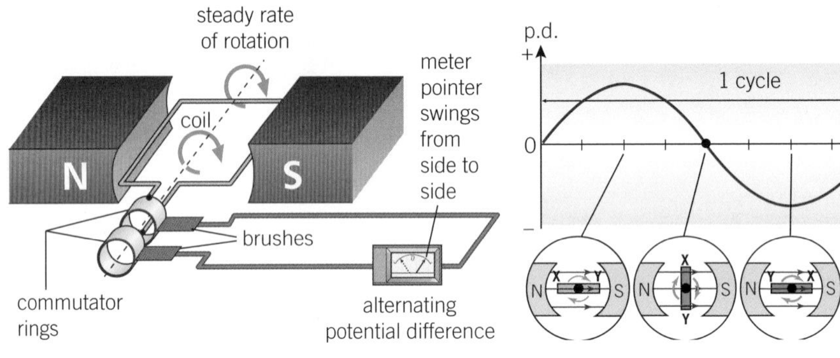

The a.c. generator

Alternating potential difference

- A **dynamo** is a direct-current generator. It is the same as an alternator except that the dynamo has a split-ring commutator instead of slip rings. A dynamo always generates a positive potential difference.
- A moving coil microphone generates an alternating p.d. as sound waves make a coil vibrate. A moving coil loudspeaker produces sound waves when an alternating p.d. is applied across its coil.

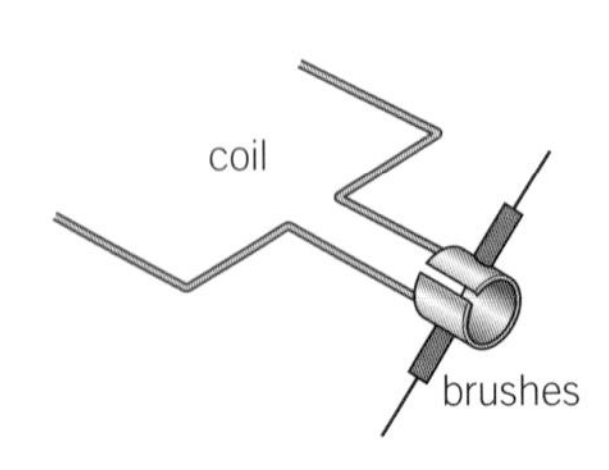

The d.c. generator

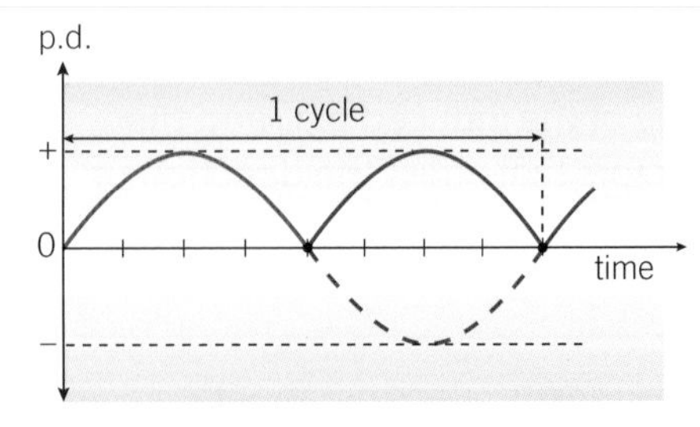

Direct current potential difference

1 What is an alternator?
2 What is a dynamo?

15.7 Transformers

Key points

- Transformers are used to increase or decrease the size of an alternating potential difference.
- The size of alternating potential difference is increased by a step-up transformer, and decreased by a step-down transformer.
- A transformer works only with a.c. because a changing magnetic field is necessary to induce a.c. in the secondary coil.
- A transformer has a primary coil, a secondary coil, and an iron core.

Synoptic link

For more information on alternating current, look at Topic P5.1.

Key word: transformer

Study tip

Write down a set of bullet points to explain how a transformer works.

On the up

You should be able to explain the structure of the transformer and how it works.

- A **transformer** consists of two coils of insulated wire, called the primary coil and the secondary coil. These coils are wound around the same iron core. When an alternating current passes through the primary coil, it produces an alternating magnetic field in the core. This field continually changes direction as the direction of current in the primary coil changes.

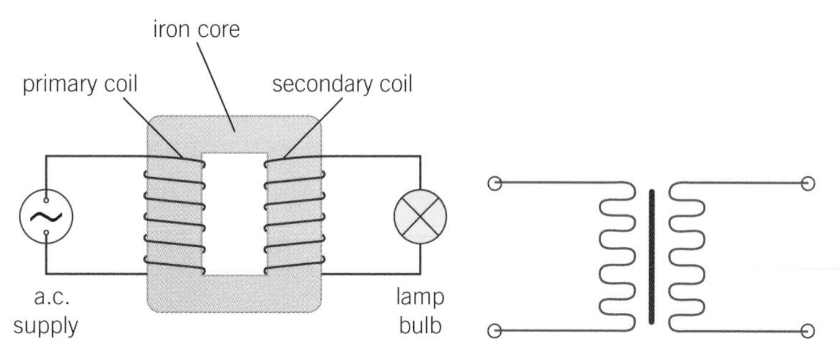

The structure of a transformer

- The alternating magnetic field lines pass through the secondary coil and induce an alternating potential difference across its ends. If the secondary coil is part of a complete circuit, an alternating current is produced.
- A transformer only works with alternating current, as a changing magnetic field across the secondary coil is needed to induce a p.d. in that coil.
- The coils of wire are insulated so that current does not short across either the iron core or adjacent turns of wire, but flows around the whole coil. The core is made of iron so it is easily magnetised and demagnetised.
- Transformers are used in the National Grid.
  - A step-up transformer makes the p.d. across the secondary coil greater than the p.d. across the primary coil. Its secondary coil has more turns than its primary coil.
  - A step-down transformer makes the p.d. across the secondary coil less than the p.d. across the primary coil. Its secondary coil has fewer turns than its primary coil.

1 Do transformers work with a.c. or d.c?

2 Why is the core of a transformer made from iron?

15.8 Transformers in action

Key points

- The transformer equation is:
  $$\frac{\text{primary p.d. } V_p}{\text{secondary p.d. } V_s} = \frac{n_p}{n_s},$$
  where n_p = number of primary turns, and n_s = number of secondary turns.
- For a step-down transformer, n_s is less than n_p.
  For a step-up transformer, n_s is greater than n_p.
- For a 100% efficient transformer:
  $V_p \times I_p = V_s \times I_s$
  where I_p = primary current, and I_s = secondary current.
- A high grid potential difference reduces the current that is needed, so it reduces power loss and makes the system more efficient.

1. What sort of transformers are used at local substations?
2. Why is a transformer used to step up the p.d. from a power station?

Synoptic link

For more information on alternating current, look at Topic P5.1.

- For a transformer, the potential difference (p.d.) across the primary and secondary coils is related to the number of turns on each coil by the equation $\frac{V_p}{V_s} = \frac{n_p}{n_s}$, where:

  V_p is the p.d. across the primary coil in volts, V
  V_s is the p.d. across the secondary coil in volts, V
  n_p is the number of turns on the primary coil
  n_s is the number of turns on the secondary coil.
- A step-down transformer has fewer turns on the secondary coil than on the primary coil.
- A step-up transformer has more turns on the secondary coil than on the primary coil.
- Most transformers are almost 100% efficient. This means that no power is lost, and so

  $V_p \times I_p = V_s \times I_s$, where
  I_p is the current in the primary coil in amperes, A
  I_s is the current in the secondary coil in amperes, A.
- In power stations, electricity is generated at a particular p.d. The p.d. is increased by step-up transformers before the electricity is transmitted across the National Grid. This reduces the current, and so reduces the heating effect and power loss in the cables, making the system more efficient. The p.d. is reduced to safe levels by step-down transformers before reaching consumers.

A step-up transformer is used to change a p.d. of 12 V to a p.d. of 240 V. If there are 50 turns on the primary coil, how many turns are there on the secondary coil?

$V_p = 12\text{ V}$ $\quad V_s = 240\text{ V}$ $\quad n_p = 50$

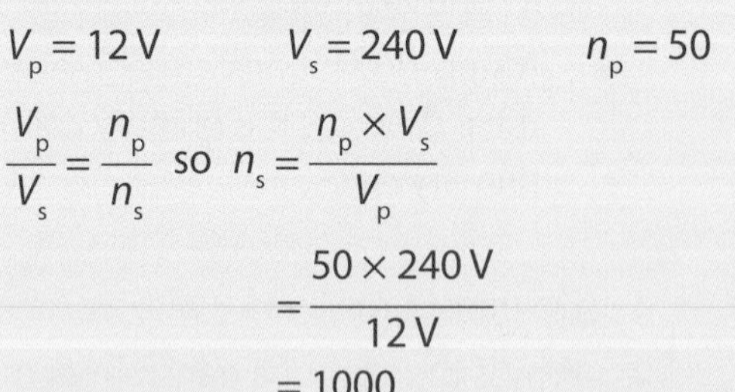

$$\frac{V_p}{V_s} = \frac{n_p}{n_s} \text{ so } n_s = \frac{n_p \times V_s}{V_p}$$
$$= \frac{50 \times 240\text{ V}}{12\text{ V}}$$
$$= 1000$$

There are 1000 turns on the secondary coil.

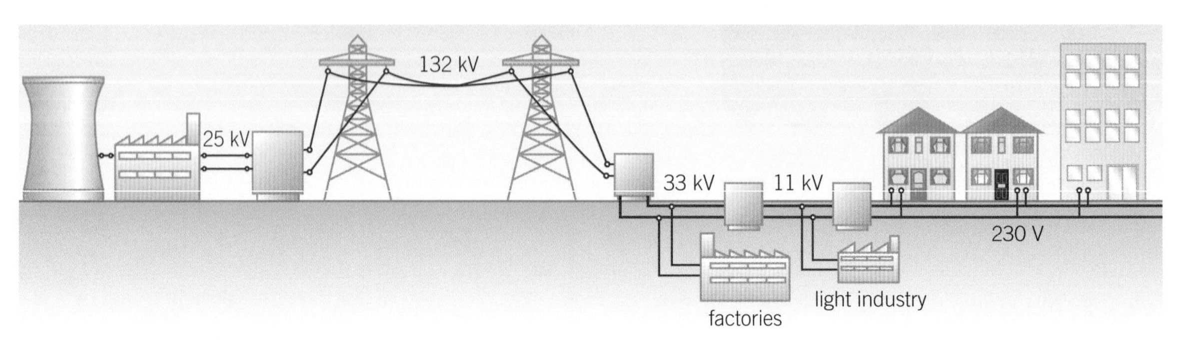

The grid system

1. Explain why steel is used to make permanent magnets? [1 mark]
2. Describe what happens if the north pole of a bar magnet is brought close to the north pole of another bar magnet. [1 mark]
3. Explain why electromagnets are useful. [2 marks]
4. Identify how the strength of the magnetic field around a solenoid be increased? [2 marks]
5. Describe the magnetic field around a bar magnet. [1 mark]
6. Explain what is meant by induced magnetism. [3 marks]
7. Describe how an electromagnet can be used in a scrapyard to separate steel cans from aluminium cans. [2 marks]
8. (H) Describe the difference between an alternator and a dynamo. [3 marks]
9. (H) Describe how Fleming's left-hand rule can be used to determine the direction of the force on a current-carrying conductor in a magnetic field. [3 marks]
10. (H) Describe the process of electromagnetic induction. [3 marks]
11. (H) Explain why the core of a transformer is made from iron. [3 marks]
12. (H) A transformer has 100 turns on the primary coil and 400 turns on the secondary coil. The p.d. across the primary coil is 2 V.
    Calculate the p.d. across the secondary coil. [3 marks]

Chapter checklist

Tick when you have:

reviewed it after your lesson	☐	☐	☐
revised once – some questions right	✓	☐	☐
revised twice – all questions right	✓	☐	✓

Move on to another topic when you have all three ticks

15.1 Magnetic fields	☐	☐	☐
15.2 Magnetic fields of electric currents	☐	☐	☐
15.3 Electromagnets in devices	☐	☐	☐
15.4 The motor effect	☐	☐	☐
15.5 The generator effect	☐	☐	☐
15.6 The alternating-current generator	☐	☐	☐
15.7 Transformers	☐	☐	☐
15.8 Transformers in action	☐	☐	☐

16.1 Formation of the Solar System

Key points

- The Solar System formed from gas and dust clouds that gradually became more and more concentrated because of gravitational attraction.
- A protostar is a concentration of gas and dust that becomes hot enough to cause nuclear fusion.
- Energy is released inside a star because of hydrogen nuclei fusing together to form helium nuclei.
- The Sun is stable because gravitational forces acting inwards balance the forces of nuclear fusion energy in the core acting outwards.

Synoptic link

For more information on fusion, look at Topic P7.8.

Key words: protostar, main sequence

Study tip

Draw a diagram to show the forces acting in a main sequence star.

On the up

You should be able to explain how fusion reactions lead to an equilibrium between the gravitational collapse of a star and the expansion of a star due to fusion energy.

- Gravitational forces pull clouds of dust and gas together to form a **protostar**.
- The protostar becomes denser and the nuclei of hydrogen atoms and other light elements start to fuse together. Energy is released in the process, so the core gets hotter and brighter.
- Stars radiate energy because of hydrogen fusion in the core. This stage can continue for billions of years until the star runs out of hydrogen nuclei in its core.
- The star is stable because the inward force of gravity is balanced by the outward force of radiation from the core and is called a **main sequence** star.
- Objects that are too small to become stars can be attracted by a protostar to become planets orbiting the star. The Solar System consists of the Sun; its orbiting planets; and other, smaller, objects, such as moons, comets, asteroids, and meteors.

1 What process releases energy in stars?

2 What is the name of the first stage in the life cycle of a star?

The Solar System

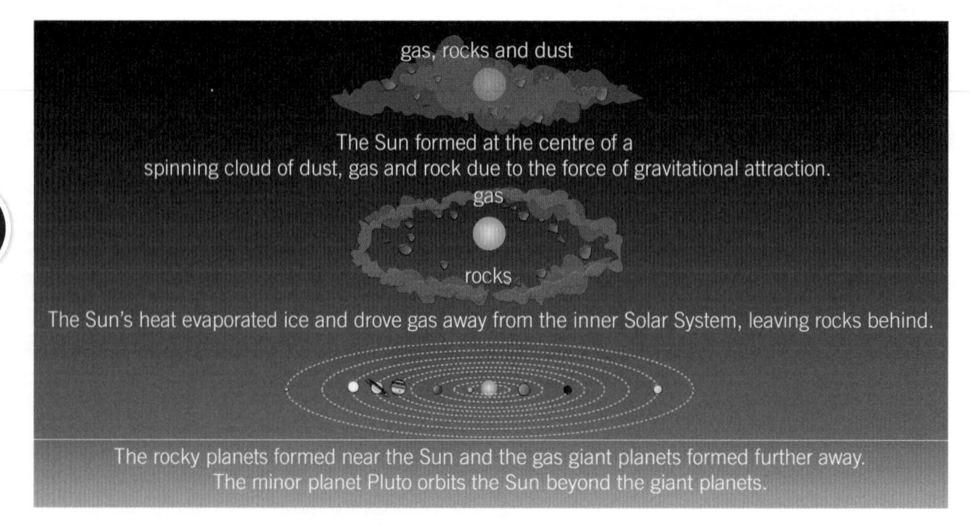

The formation of the Solar System. The planets move round the Sun in circular or almost-circular orbits that are in the same plane as each other. The Earth is the third planet from the Sun. Its orbit is in the 'habitable' zone round the Sun where the temperature is between 0 °C and 100 °C, so liquid water can exist on a planet there

16.2 The life history of a star

Key points

- Stars become unstable when they have no more hydrogen nuclei that they can fuse together.
- Stars with about the same mass as the Sun: protostar → main sequence star → red giant → white dwarf → black dwarf.
- Stars much more massive than the Sun: protostar → main sequence star → red supergiant → supernova → neutron star (or black hole if enough mass).
- The Sun will eventually become a black dwarf.
- A supernova is the explosion of a red supergiant after it collapses.

Synoptic link

For more information on fusion, look at Topic P7.8.

Study tip

Draw a labelled diagram to help you remember the stages in the life cycle of a star.

- A star on the main sequence will eventually run out of hydrogen in its core. It will swell, cool down, and turn red.
- A star of similar size to the Sun will become a **red giant**. At this stage, helium and other light elements in the core fuse to become heavier elements. Fusion stops and the star will contract to form a **white dwarf**. Eventually no more light is emitted and the star becomes a **black dwarf**.
- A star much larger than the Sun will swell to become a **red supergiant**, which then collapses. Eventually the star explodes in a **supernova**.
- Elements heavier than iron are only formed in the final stages of the life of a big star. This is because the process requires the input of so much energy. All the elements get distributed through space by the supernova explosion.
- During the supernova the outer layers are thrown into space. The core is left as a **neutron star**. If this is massive enough it becomes a **black hole**. The gravitational field of a black hole is so strong not even light can escape from it.

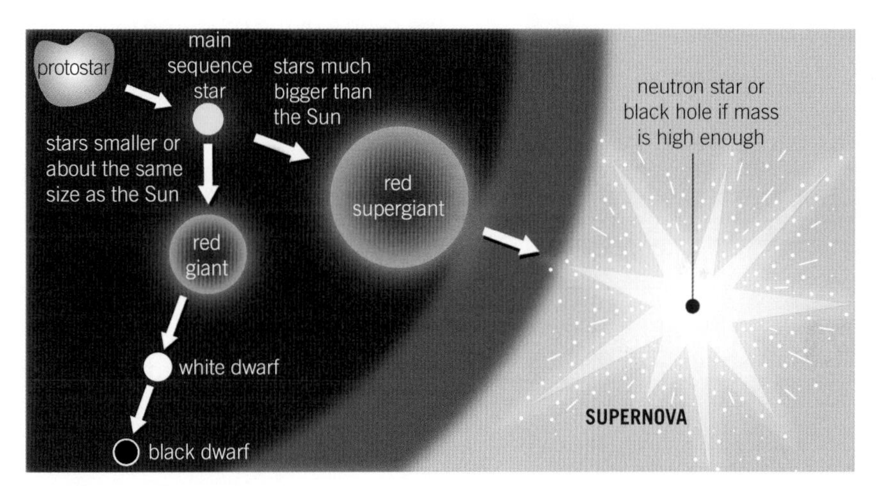

The life cycle of a star

1 What is the heaviest element formed by fusion in a main sequence star?
2 How are the chemical elements distributed through space?

Key words: red giant, white dwarf, black dwarf, red supergiant, supernova, neutron star, black hole

16.3 Planets, satellites, and orbits

Key points

- The force of gravity between:
  - a planet and the Sun keeps the planet moving along its orbit
  - a satellite and the Earth keeps the satellite moving along its orbit.
- (H) The force of gravity on an orbiting body in a circular orbit is towards the centre of the circle.
- As a body in a circular orbit moves around the orbit:
  - the magnitude of its velocity (its speed) does not change
  - the direction of its velocity continually changes and is always at right angles to the direction of the force, so
  - (H) it experiences an acceleration towards the centre of the circle.
- (H) To stay in orbit at a particular distance, a small body must move at a particular speed around a larger body.

Synoptic links

For more information on resultant forces, look at Topic P8.3.

For more information on force and acceleration, look at Topic P10.1.

Key word: centripetal force

- The Earth moves around the Sun in a path called an orbit. The other planets orbit the Sun, and the Moon and artificial satellites orbit the Earth. The orbits are circular or almost circular. In each case a smaller object orbits a much bigger object.
- A force is needed to keep the smaller object moving along its orbit. The force is the force of gravity between the bigger object and the smaller object.

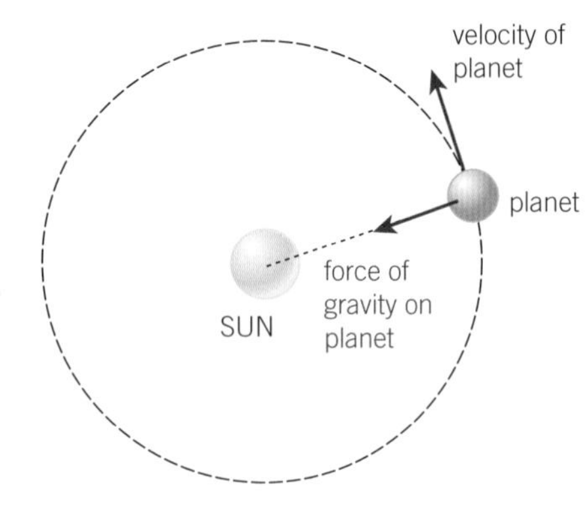

A circular orbit

- As an object moves around an orbit, its speed is constant but its direction continually changes, so its velocity continually changes. It accelerates towards the centre of the orbit. The direction of the velocity is always at right angles to the direction of the gravitational force on it. The gravitational force is always towards the centre of the bigger object.

1 What is the name of the force between the Earth and the Sun?

Higher

The force needed to keep an object moving in a circle is called the **centripetal force**.

This force always acts towards the centre of the circle.

For a planet or satellite in an orbit around a bigger object, the centripetal force is the gravitational force from the bigger object and acts towards its centre.

The centripetal force is a resultant force, so it produces an acceleration. The acceleration is towards the centre of the orbit.

For an object to orbit at a particular distance from the bigger object it must go at a particular speed. If the distance decreases, the speed will increase.

For an object to remain in the Earth's orbit, it must travel at a particular speed. Too great a speed and the object will fly off into space. Too small a speed and the object will fall down to Earth.

2 What is the name of the force that keeps an object moving in a circle?

Student Book pages 238–239

16.4 The expanding universe

Key points

- The red-shift of a distant galaxy is the shift to longer wavelengths (and lower frequencies) of the light from the galaxy because it is moving away from you.
- The faster a distant galaxy is moving away from you, the greater its red-shift is.
- All the distant galaxies show a red-shift. The further away a distant galaxy is from you, the greater its red-shift is.
- The distant galaxies are all moving away from you because the universe is expanding.

Synoptic link

For more information on wavelength and frequency of waves, look at Topic P12.2.

Key word: red-shift

On the up

To achieve the top grades, you should be able to explain how observed red-shift provides evidence that the universe is expanding and supports the Big Bang theory.

- Galaxies are large collections of stars.

The Milky Way galaxy

- Light observed from distant galaxies undergoes **red-shift**. The frequency has decreased and the wavelength has increased. The spectrum of light is shifted towards the red part of the spectrum. Red-shift shows that the galaxy is moving away from us.
- A blue shift would indicate that a galaxy is moving towards us. We are able to see these effects by observing dark lines in the spectra from galaxies.

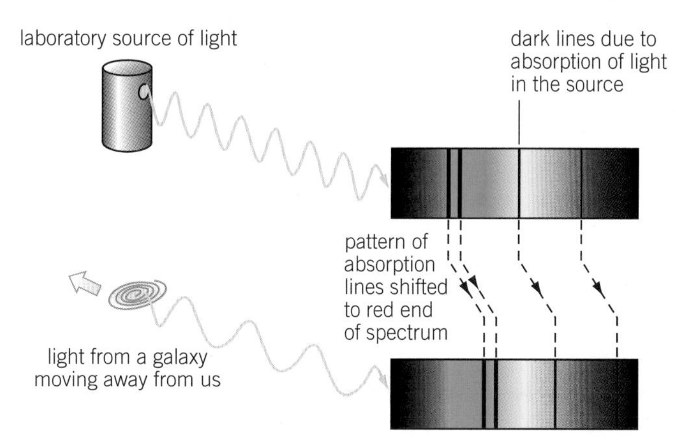

Red-shift

- The further away the galaxy, the bigger the red-shift. This suggests that distant galaxies are moving away from us and that the most distant galaxies are moving the fastest. This is true of galaxies no matter which direction you look in.
- All the distant galaxies are moving away from each other, so the whole universe is expanding.

1 Which galaxies are moving fastest?

2 What happens to the wavelength of light that is red-shifted?

Student Book pages 240–241 P16

16.5 The beginning and future of the Universe

Key points

- The universe started with the Big Bang, which was a massive explosion from a very small point.
- The universe has been expanding ever since the Big Bang.
- Cosmic microwave background radiation (CMBR) is electromagnetic radiation that was created just after the Big Bang.
- The red-shifts of the distant galaxies provide evidence that the universe is expanding. CMBR can be explained only by the Big Bang theory.

Synoptic link

For more information on gamma radiation, look at Topic P13.4.

Key words: Big Bang, cosmic microwave background radiation (CMBR), dark matter

- Red-shift gives us evidence that the universe is expanding outwards in all directions.
- We can imagine going back in time to see where the universe came from. If it is now expanding outwards, this suggests that it started with a massive explosion at a very small initial point. This is known as the **Big Bang** theory.
- If the universe began with a Big Bang, then high-energy gamma radiation would have been produced. As the universe expanded, this would have become lower-energy radiation.
- Scientists discovered microwaves coming from every direction in space. This is **cosmic microwave background radiation (CMBR)**, the radiation produced by the Big Bang.
- The Big Bang theory is so far the only way to explain the existence of CMBR.
- What will ultimately happen to the universe depends on the density of the universe.
- The stars in a galaxy are only a small percentage of its total mass. The missing mass is called **dark matter**. The average density of the universe is much bigger than it would be if dark matter didn't exist.
- If the density of the universe is less than a particular amount, it will expand forever – the Big Yawn.
- If the density of the universe is more than a particular amount, it will stop expanding and go into reverse – the Big Crunch.

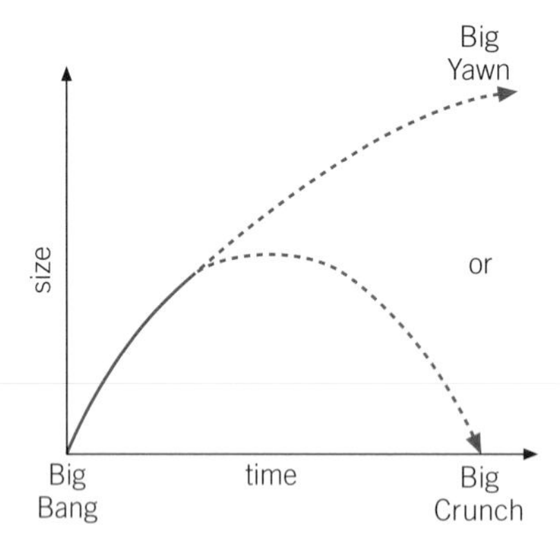

The future of the universe?

1. What is the Big Bang theory?
2. What happens to the wavelength of radiation as it changes from a gamma ray to a microwave?

1. What is an orbit? [1 mark]
2. Describe what is meant by a black hole. [2 marks]
3. What force keeps the Moon in orbit around the Earth? [1 mark]
4. Describe what happens to the wavelength and frequency of light that is blue-shifted. [2 marks]
5. What happens to a star much larger than the Sun as it leaves the main sequence? [1 mark]
6. Name the stages in the life cycle of a star about the same size as the Sun. [3 marks]
7. What is the relationship between the relative speed of a galaxy and the red-shift of light from the galaxy? [1 mark]
8. Explain what is meant by the Big Bang theory. [3 marks]
9. Explain why a star on the main sequence is stable. [2 marks]
10. Describe what is meant by a centripetal force. [2 marks]
11. Describe how astronomers determine if light from a galaxy is red-shifted. [3 marks]
12. Explain what is meant by cosmic microwave background radiation. [3 marks]

Chapter checklist

Tick when you have:

reviewed it after your lesson	✓		
revised once – some questions right	✓	✓	
revised twice – all questions right	✓	✓	✓

Move on to another topic when you have all three ticks

16.1 Formation of the Solar System			
16.2 The life history of a star			
16.3 Planets, satellites, and orbits			
16.4 The expanding universe			
16.5 The beginning and future of the Universe			

01 Two students carried out an experiment to measure the speed of sound. They stood on opposite sides of the school field, so they were 280 m from each other. One student held up a starting pistol and fired it into the air. The second student started a stopwatch when he saw the smoke from the starting pistol and stopped the stopwatch when he heard the bang from the starting pistol. They repeated this twice more. Their recorded times were 0.85 s, 0.90 s, and 0.80 s.

01.1 Explain why the students stood as far away from each other as possible. [2 marks]

01.2 Explain why the students repeated the experiment three times. [2 marks]

01.3 Calculate an estimate for the speed of sound in air using the students' results. [3 marks]

02 A student investigated the amount of infrared radiation emitted by different surfaces. She filled two cans with hot water. One can had been painted matt black and the other can had been painted shiny silver. She used a thermometer to measure the temperature of the water in each can. She left both cans on a metal bench for five minutes, then measured the temperature of the water in each can again.

02.1 In addition to leaving both cans to cool for the same time before measuring the temperature again, give two other control variables the student should keep the same in this investigation. [2 marks]

02.2 Suggest two improvements to the student's method. [2 marks]

02.3 Complete the sentence:
The can with the ______________ temperature at the end of the investigation is the best emitter of infrared radiation. [1 mark]

03 A student makes an electromagnet from an iron nail, a length of wire, and a battery, as shown in **Figure 1**.

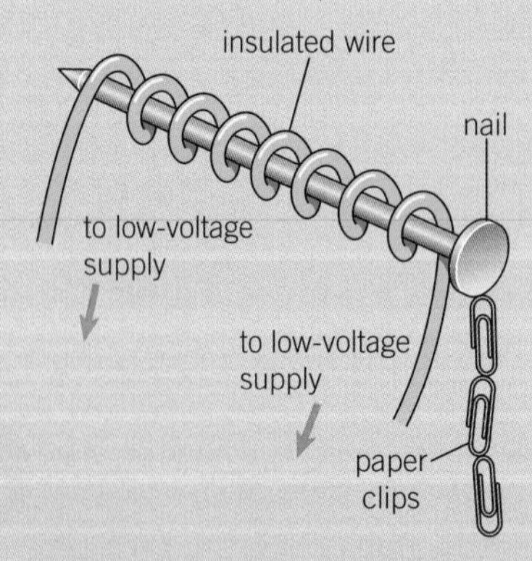

Figure 1

When the current is switched on the student uses the electromagnet to pick up some paperclips.

03.1 Design an investigation the student can carry out to test how the strength of the electromagnet varies with the number of turns of wire around the nail. [3 marks]

03.2 Explain what will happen if the student repeats the investigation with a greater current in the wire. [3 marks]

03.3 Explain what will happen if the student repeats the investigation with a wooden rod in place of the nail. [3 marks]

04 Many objects reflect light.

04.1 Explain the difference between specular reflection and diffuse reflection. [5 marks]

04.2 **Figure 2** shows two rays of light from a pin, incident on a mirror.

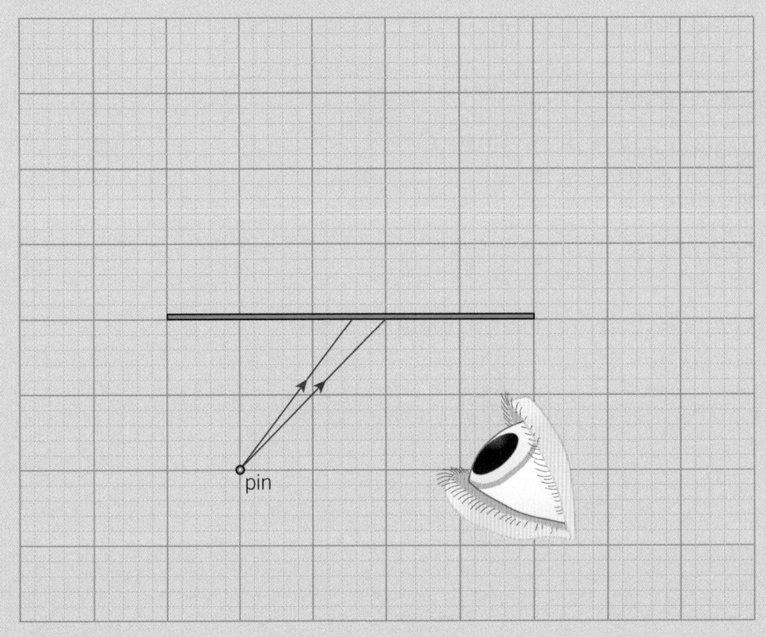

Figure 2

Copy and complete the diagram to show the paths of the two rays to the eye and use them to locate the position of the image of the pin. [4 marks]

Study tip

Draw the path of the rays carefully with a ruler and make sure you show the reflection at the exact point the incident ray meets the mirror.

05 (H) A laptop requires a 20V supply to operate. It is powered by a transformer which has 11 500 turns on the primary coil.

05.1 What is the core of a transformer made from? [1 mark]

05.2 The potential difference of the alternating mains supply is 230V. Calculate the number of turns on the secondary coil of the transformer. [3 marks]

05.3 Explain why a transformer will only work with an alternating supply. [2 marks]

Study tip

The transformer equation is tricky. Take care when rearranging it.

06 Loudspeakers produce sound waves.

06.1 Sound waves are longitudinal waves. Explain how sound waves move through the air as longitudinal waves. [3 marks]

06.2 (H) **Figure 3** shows a moving coil loudspeaker.
The loudspeaker contains a moveable coil attached to a diaphragm. The diaphragm fits loosely over a cylindrical permanent magnet. An amplifier produces a varying, alternating current in the coil.
Explain how the loudspeaker makes use of the motor effect to produce a sound wave. [6 marks]

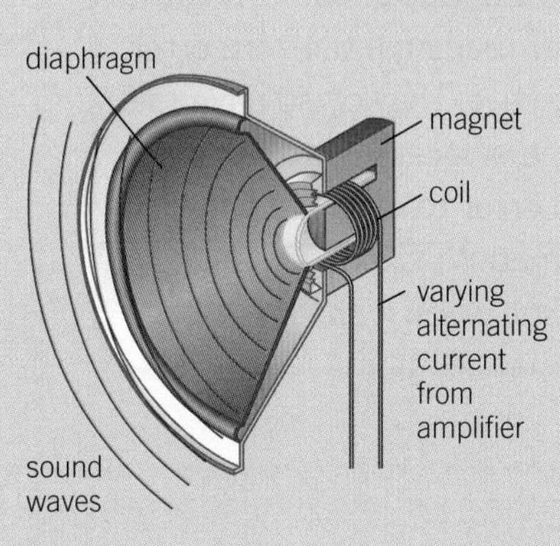

Figure 3

Introduction

This is an example of a question focusing on practical skills. The student answers are marked and include comments to help you answer this type of question as effectively as possible.

The correct control variables have been ticked.

01 Some students were asked to investigate the factors that determine the resistance of a piece of wire.

One group of students decided to find out how the resistance depends on the length of the wire.

01.1 Choose three of the following variables that the students should control.

Tick three boxes.

colour of wire	
diameter of wire	✓
length of wire	
temperature of wire	✓
type of wire	✓

[3 marks]

01.2 The students set up the circuit in **Figure 1** below.

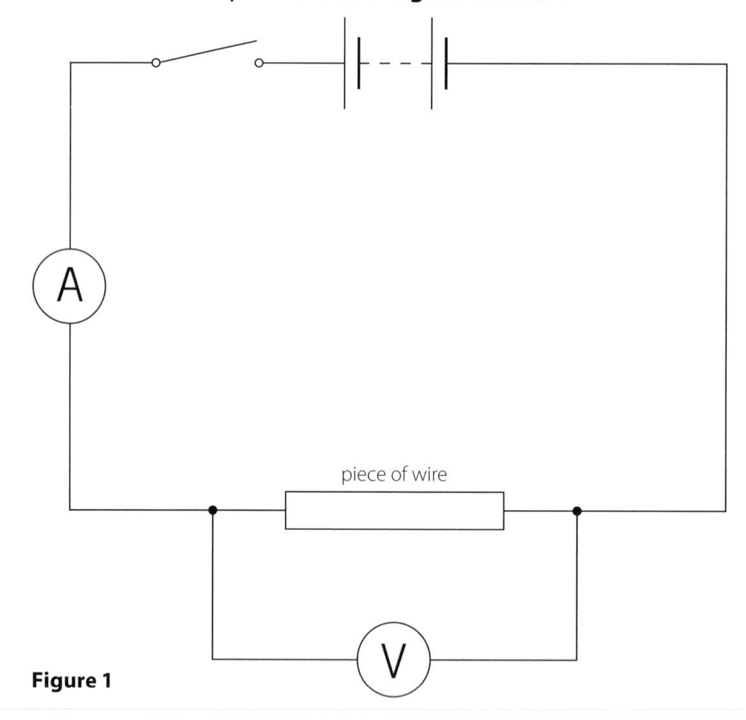

Figure 1

One of the meters had a zero error.

Explain what is meant by a zero error and how you would correct readings taken from the meter with the zero error. [2 marks]

A zero error is when the pointer on a meter does not return to 0 when there is no current. If the meter has a reading when it should read zero, this value should be subtracted from every reading.

The student has correctly stated what a zero error is and given the way of correcting the meter reading if the zero error is positive. They could have said that it is necessary to add the zero error to each reading if the meter has a negative reading when it should read 0.

01.3 The students cut six different lengths from a reel of wire. They measured the length of each piece of wire.

Then they used crocodile clips to connect a length of wire in the circuit, as shown in **Figure 2**.

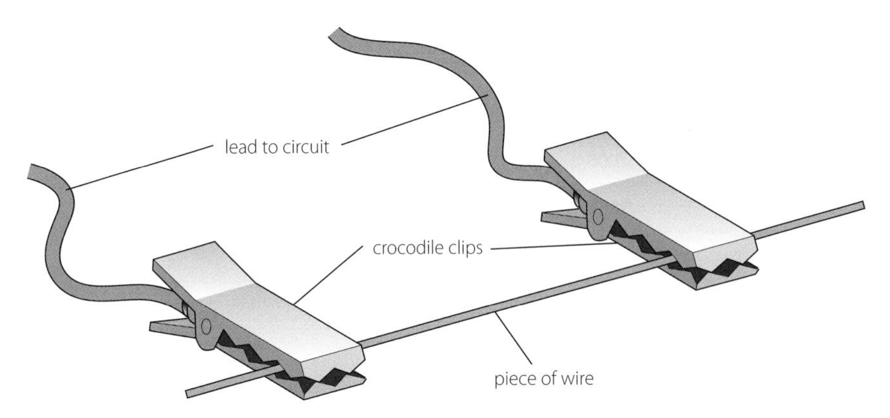

Figure 2

Identify a fault in their technique and suggest how the measurement of length could be improved. [2 marks]

The length of wire in the circuit is less than the measured length. The measurement could be improved by measuring the distance between the points of contact of the crocodile clips with the wire.

The student has given the correct answer – this is a case where a diagram, marking exactly what should be measured, would help.

01.4 Suggest and explain what precaution the students should take when using short lengths of wire, and why they should not use pieces less than 5 cm long. [5 marks]

They should turn the switch off between readings. Short lengths of wire will have low resistance, so there will be a large current and the wire will get hot. Lengths less than 5 cm might get extremely hot and even melt.

The student would be awarded five marks, since they have made five relevant points. Other points could gain credit: if the wire gets hot, one of the control variables has not been kept constant and the results will not be valid; the percentage error in measuring length will be higher for shorter wires: 1 mm in 5 cm = 2%; 1 mm in 1 m = 0.1%.

01.5 The students obtained the data they needed and repeated the procedure with the other pieces of wire. They then calculated the resistance of each piece of wire.

Their results are given in the table.

Length of wire in cm	Resistance in Ω
40	0.8
50.00	1.0
60.0	1.2
70	0.4
80.0	1.6
90	1.8

What equation should the students use to calculate resistance? [1 mark]

R = V ÷ I

Some students may put $V = IR$ or put the equation in words rather than letters and still score the mark.

01.6 Explain what is wrong with the recorded readings of length of wire in the table. [2 marks]

The readings are inconsistent. A normal ruler can measure to the nearest mm so all results should be recorded to the first decimal place.

The answer could have been improved by describing the inconsistencies – some lengths to the nearest cm, others to the nearest mm, and one to 0.01 cm, but both marks would be scored.

01.7 The results have been plotted on the grid below.
Draw a line of best fit. [2 marks]

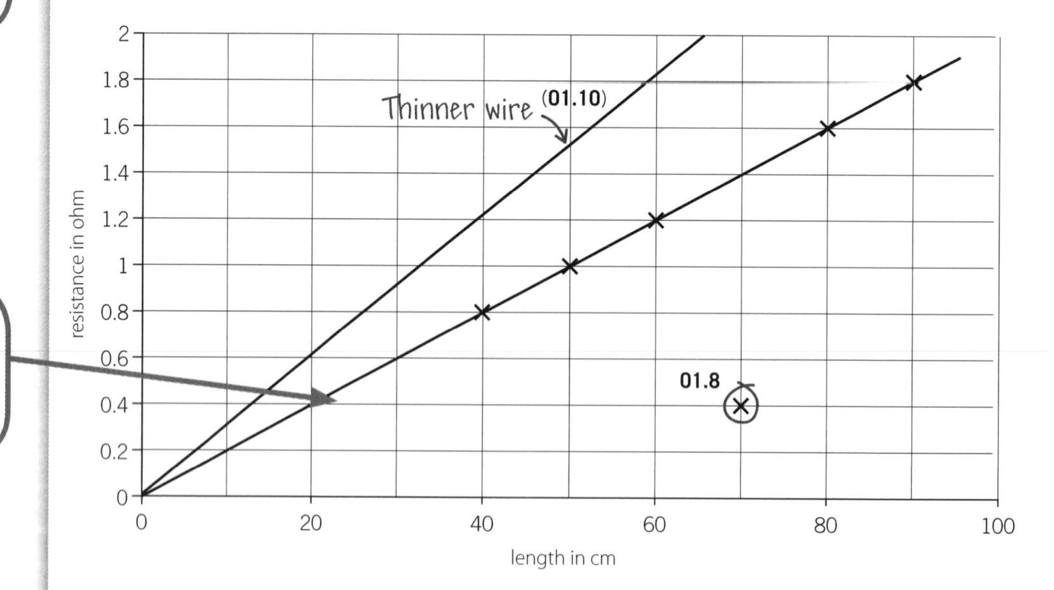

A straight line, through (0, 0), has been drawn with the aid of a ruler and the anomalous result has been ignored.

01.8 One of the results is anomalous.
Draw a ring around the anomalous result on the graph. [1 mark]

Ring round (70, 0.4)

The anomalous point has been correctly identified.

01.9 Describe the relationship between the length of a wire and its resistance. [2 marks]

As the length is doubled, the resistance is doubled

The student could have said that the two quantities are directly proportional. This is indicated by the straight line through the origin of the graph.

01.10 Another group of students carried out the same investigation. They used a reel of thinner wire.

Draw a line on the graph to show how their results were different from those of the first group of students. Label this line 'Thinner wire'. [1 mark]

Graph line with greater gradient

Thinner wires have a higher resistance per unit length so the line has a greater slope.

01 In an experiment to measure the speed of water waves in a ripple tank, a student obtained the following measurements of the time taken by a water wave to travel the length of the tank ten times.

26.2 s, 25.8 s, 26.8 s, 27.2 s, 26.4 s

01.1 Calculate the mean value of the time taken for the wave to travel the length of the tank ten times. [1 mark]

01.2 The length of the tank was 0.440 m. Calculate the speed of the waves in the tank. [2 marks]

01.3 The precision of the stopwatch used by the student was 0.01 s. Give one reason why the measurements above are not the same. [1 mark]

02 In a low-voltage lighting circuit, two lamps A and B are connected in parallel to a 12 V battery, as shown in **Figure 1**. Each lamp is switched on or off using a switch in series with it.

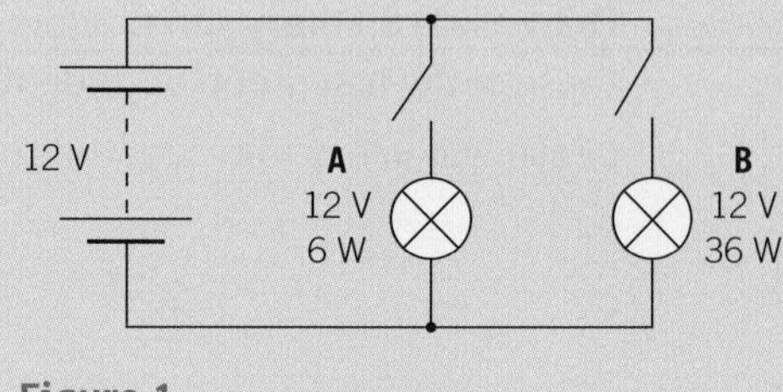

Figure 1

02.1 Lamp **A** is a 12 V 6 W lamp. Lamp **B** is a 12 V 36 W lamp.

Calculate the current through each lamp when it is switched on. [2 marks]

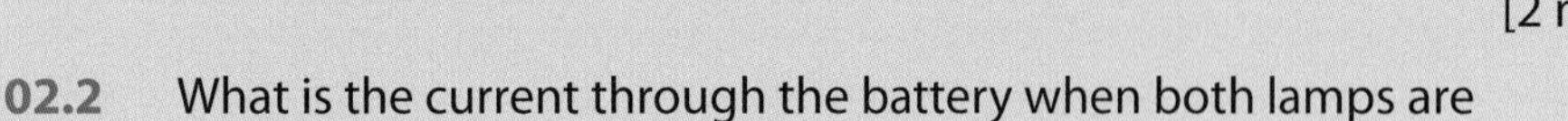

02.2 What is the current through the battery when both lamps are switched on? [1 mark]

02.3 A student suggests the two lamps could be connected in series with the battery and one switch.

Give one reason why this series circuit would be less useful than the parallel circuit. [1 mark]

02.4 Compare the brightnesses of the lamps in the parallel circuit shown in **Figure 1** and in the suggested series circuit. [3 marks]

03 A student carried out an experiment to measure the spring constant of a new spring. She carried out a preliminary test by measuring the length of the spring using a metre rule and a set square, as shown in **Figure 2**.

She measured the position of each end of the spring when it supported a 1.0 N weight hanger. She then measured the two positions when a 4.0 N weight was added to the weight hanger. She then repeated the test. **Table 1** shows the measurements she made.

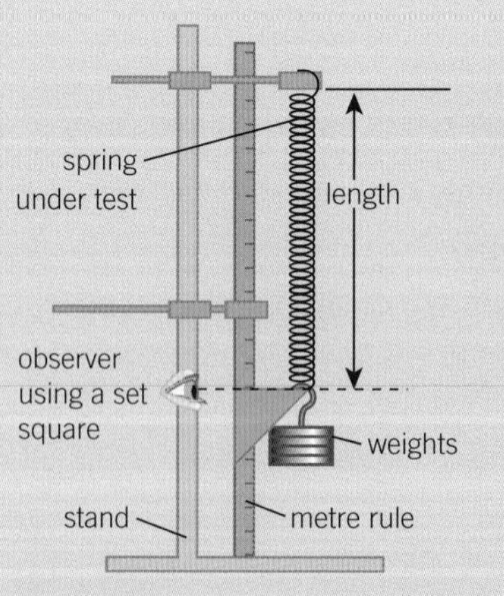

Figure 2

Table 1

Total weight in N	Position in mm		Length of spring in mm
	Top end	Lower end	
1.0	5	157	152
	5	159	154
5.0	5	315	310
	5	311	306

03.1 Calculate the mean value of the length of the spring when it supported a weight of 1.0 N. [1 mark]

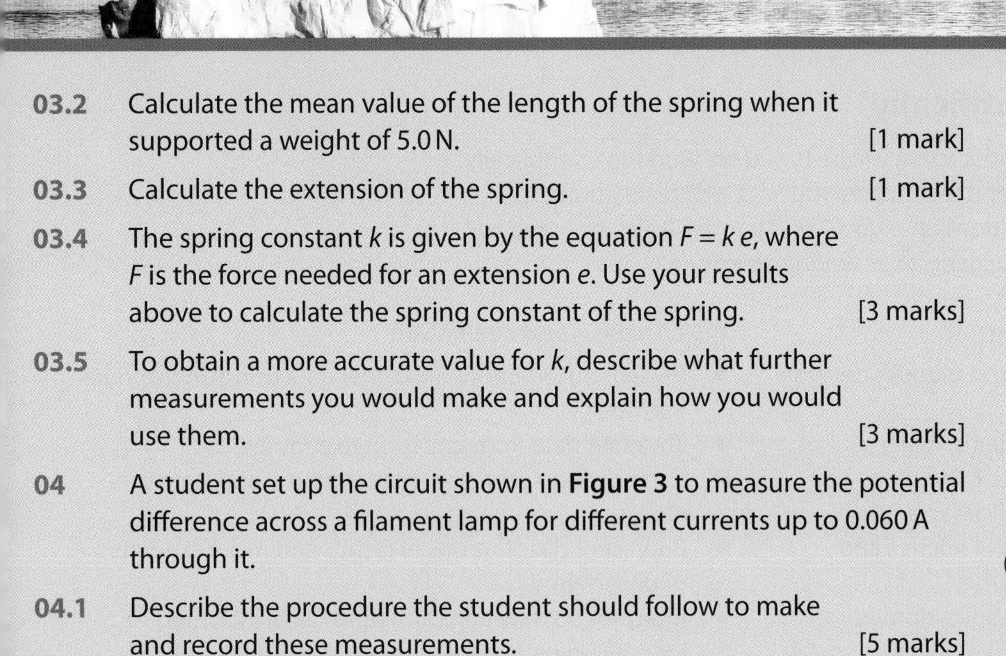

03.2 Calculate the mean value of the length of the spring when it supported a weight of 5.0 N. [1 mark]

03.3 Calculate the extension of the spring. [1 mark]

03.4 The spring constant k is given by the equation $F = k\,e$, where F is the force needed for an extension e. Use your results above to calculate the spring constant of the spring. [3 marks]

03.5 To obtain a more accurate value for k, describe what further measurements you would make and explain how you would use them. [3 marks]

04 A student set up the circuit shown in **Figure 3** to measure the potential difference across a filament lamp for different currents up to 0.060 A through it.

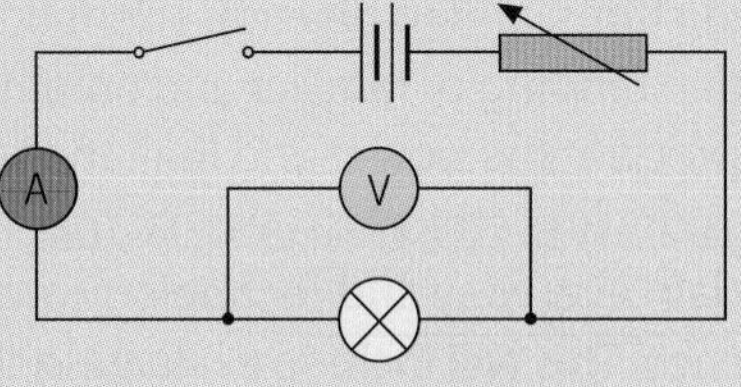

Figure 3

04.1 Describe the procedure the student should follow to make and record these measurements. [5 marks]

04.2 **Table 2** shows a set of results obtained by the student.

Table 2

Current in A	0	0.010	0.021	0.032	0.042	0.052	0.057
Potential difference in V	0	0. 60	1.40	2.35	3.50	4.90	5.90

Plot a graph of these measurements. [3 marks]

04.3 Calculate the resistance of the lamp at 3.0 V and at 6.0 V. [2 marks]

04.4 Explain in terms of electrons why the resistance of the lamp is greater at 6.0 V than at 3.0 V. [3 marks]

04.5 The battery in the circuit is a rechargeable battery that can supply a current of 0.06 A for a maximum of 10 hours.

Calculate the charge that would flow through the battery when the current is 0.06 A for 10 hours. [2 marks]

04.6 Calculate the energy that would be supplied to a lamp operating at 6.0 V when this amount of charge passes through it. [2 marks]

Summary of 'Working scientifically'

A minimum of 15% of the total marks in your exam will be based on 'Working scientifically'. Questions could ask about the methods and techniques you have practised during the Required practicals (see the Practice questions on pages 126–7). You will also be asked to apply the principles of scientific enquiry in general, as outlined below.

WS1 Development of scientific thinking

- Understand how scientific methods and theories develop over time.
- Use a variety of models to solve problems, make predictions, and to develop scientific explanations and understanding of familiar and unfamiliar facts.
- Appreciate the power and limitations of science and consider any ethical issues that may arise.
- Explain everyday and technological applications of science. This will include evaluating their personal, social, economic, and environmental implications. You may also be asked to make decisions based on the evaluation of evidence and arguments.
- Evaluate risks both in practical science and in wider contexts, including perception of risk in relation to data and consequences.
- Recognise the importance of peer review of results and of communicating results to a range of audiences.

WS2 Experimental skills and strategies

- Use scientific theories and explanations to develop hypotheses.
- Plan experiments or devise procedures to make observations, produce or characterise a substance, test hypotheses, check data, or explore phenomena.
- Apply knowledge of a range of techniques, instruments, apparatus, and materials to select those appropriate to the experiment.
- Carry out experiments, handling apparatus correctly and taking into account the accuracy of measurements, as well as any health and safety considerations.
- Recognise when to apply knowledge of sampling techniques to make sure that any samples collected are representative.
- Make and record observations and measurements using a range of apparatus and methods.
- Evaluate methods and suggest possible improvements and further investigations.

WS3 Analysis and evaluation

- Present observations and other data using appropriate methods.
- Translate data from one form to another.
- Carry out and represent mathematical and statistical analysis.
- Represent distributions of results and make estimates of uncertainty.
- Interpret observations and other data, including identifying patterns and trends, making inferences, and drawing conclusions.
- Present reasoned explanations, including relating data to hypotheses.
- Be objective; evaluating data in terms of accuracy, precision, repeatability, and reproducibility; and identifying potential sources of random and systematic error.
- Communicate the reasoning for investigations, the methods used, the findings, and reasoned conclusions through paper-based and electronic reports, as well as using other forms of presentation.

WS4 Scientific vocabulary, quantities, units, symbols, and nomenclature

- Use scientific vocabulary, terminology, and definitions.
- Recognise the importance of scientific quantities and understand how they are determined.
- Use SI units (e.g., kg, g, mg; km, m, mm; kJ, J) and IUPAC chemical names, whenever appropriate.
- Use prefixes and powers of ten for orders of magnitude (e.g., tera, giga, mega, kilo, centi, milli, micro, and nano).
- Interconvert units.
- Use an appropriate number of significant figures in calculation. Quote your answer to the same number of significant figures as the data provided in the question, to the least number of significant figures.

Essential 'Working scientifically' terms

Your knowledge of what it means to 'work scientifically' will be tested in your exams. In order to understand the nature of science and experimentation, you will need to know some technical terms that scientists use. You should be able to recognise and use the terms below:

accurate a measurement is considered accurate if it is judged to be close to the true value
anomalies/anomalous results results that do not match the pattern seen in the other data collected or that are well outside the range of other repeat readings (outliers)
categoric variable has values that are labels (described in words), for example, types of material
control variable a variable that may, in addition to the variable under investigation (the independent variable – see below), affect the outcome of an investigation and therefore has to be kept constant, or at least has to be monitored as carefully as possible
data information, either qualitative (descriptive) or quantitative (measured), that has been collected
dependent variable the variable for which the value is measured for each and every change in the variable under investigation (called the independent variable – see below)
directly proportional a relationship that, when drawn on a line graph, shows a positive linear relationship (a straight line) that crosses through the origin
fair test a test in which only the independent variable has been allowed to affect the dependent variable
gradient (of a straight line graph) change of the quantity plotted on the y-axis divided by the change of the quantity plotted on the x-axis
hazards anything that can cause harm, for example, an object, a property of a substance, or an activity
hypothesis a proposal intended to explain certain facts or observations
independent variable the variable under investigation, for which values are changed or selected by the investigator
line graph used when both variables (x and y) plotted on a graph are continuous. The line should normally be a line of best fit, and may be straight or a smooth curve

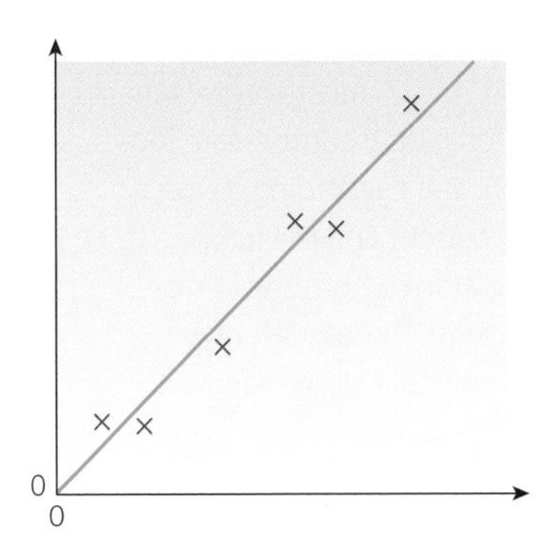

When a straight line of best fit goes through the origin (0, 0) the relationship between the variables is directly proportional

peer review evaluation of scientific research before its publication by others working in the same field
precise a precise measurement is one in which there is very little spread about the mean value. Precision depends only on the extent of random errors – it gives no indication of how close results are to the true (accurate) value
precision a measure of how precise a measurement is
prediction a forecast or statement about the way in which something will happen in the future, which is often quantitative and based on a theory or model
random error an error in measurement caused by factors that vary from one measurement to another
range the maximum and minimum values for the independent or dependent variables – this is important in ensuring that any patterns are detected and are valid
relationship the link between the variables (the independent and dependent variables, *x* and *y*) that were investigated
repeatable a measurement is repeatable if the original experimenter repeats the investigation using the same method and equipment and obtains the same results or results that show close agreement
reproducible a measurement is reproducible if the investigation is repeated by another person using different equipment and the same results are obtained
resolution this is the smallest change in the quantity being measured by a measuring instrument that gives a perceptible change in the reading
risk depends on both the likelihood of exposure to a hazard and the seriousness of any resulting harm
systematic error error that causes readings to be spread about some value other than the true value, due to results differing from the true value by a consistent amount each time a measurement is made. Sources of systematic error can include the environment, methods of observation, or instruments used. Systematic errors cannot be dealt with by simple repeats. If a systematic error is suspected, the data collection should be repeated using a different technique or a different set of equipment, and the results compared
tangent a straight line drawn to touch a point on a curve so that it has the same gradient as the curve at that point
validity the suitability of the investigative procedure to answer the question being asked
uncertainty the degree of variability that can be expected in a measurement. A reasonable estimate of the uncertainty in a mean value calculated from a set of repeat readings would be half of the range of the repeats. In an individual measurement, its uncertainty can be taken as half the smallest scale division marked on the measuring instrument or half the last figure shown on the display on a digital measuring instrument
variables physical, chemical, or biological quantities or characteristics

Glossary

Acceleration Change of velocity per second (in metres per second per second, m/s^2).

Activity The number of unstable atoms that decay per second in a radioactive source.

Alpha radiation Alpha particles, each composed of two protons and two neutrons, emitted by unstable nuclei.

Alternating current Electric current in a circuit that repeatedly reverses its direction.

Alternator An alternating current generator.

Amplitude The height of a wave crest or trough of a transverse wave from the rest position. For oscillating motion, the amplitude is the maximum distance moved by an oscillating object from its equilibrium position.

Angle of incidence Angle between the incident ray and the normal.

Angle of reflection Angle between the reflected ray and the normal.

Atomic number The number of protons (which equals the number of electrons) in an atom. It is sometimes called the proton number.

Beta radiation Beta particles that are high-energy electrons created in, and emitted from, unstable nuclei.

Big Bang theory The theory that the universe was created in a massive explosion (the Big Bang), and that the universe has been expanding ever since.

Biofuel Any fuel taken from living or recently living materials, such as animal waste.

Black body radiation The radiation emitted by a perfect black body (a body that absorbs all the radiation that hits it).

Black dwarf A star that has faded out and gone cold.

Black hole An object in space that has so much mass that nothing, not even light, can escape from its gravitational field.

Boiling point Temperature at which a pure substance boils or condenses.

Boyle's Law For a fixed mass of gas at constant temperature, its pressure multiplied by its volume is constant.

Braking distance The distance travelled by a vehicle during the time it takes for its brakes to act.

Carbon-neutral Describes a biofuel from a living organism that takes in as much carbon dioxide from the atmosphere as is released when the fuel is burnt.

Carrier waves Waves used to carry any type of signal.

Centripetal force The resultant force towards the centre of a circle acting on an object moving in a circular path.

Chain reaction A reaction in which one reaction causes further reactions, which in turn cause further reactions, etc.

Charge-coupled device (CCD) An electronic device that creates an electronic signal from an optical image formed on the CCD's array of pixels.

Closed system An object or a group of objects for which the total energy is constant.

Compression Squeezing together.

Concave (diverging) lens A lens that makes light rays parallel to the axis diverge (spread out) as if from a single point.

Conservation of energy Energy cannot be created or destroyed.

Conservation of momentum In a closed system, the total momentum before an event is equal to the total momentum after the event. Momentum is conserved in any collision or explosion, provided no external forces act on the objects that collide or explode.

Contrast medium An X-ray-absorbing substance used to fill a body organ so the organ can be seen on a radiograph.

Convex (converging) lens A lens that makes light rays parallel to the principal axis converge (meet) at a point.

Cosmic microwave background radiation (CMBR) Electromagnetic radiation that has been travelling through space ever since it was created shortly after the Big Bang.

Count rate The number of counts per second detected by a Geiger counter.

Dark matter Matter in a galaxy that cannot be seen. Its presence is deduced because galaxies would spin much faster if their stars were their only matter.

Deceleration Change of velocity per second when an object slows down.

Density Mass per unit volume of a substance.

Diffuse reflection Reflection from a rough surface – the light rays are scattered in different directions.

Diode A non-ohmic conductor that has a much higher resistance in one direction (its reverse direction) than in the other direction (its forward direction).

Direct current Electric current in a circuit that is in one direction only.

Directly proportional A graph will show this if the line of best fit is a straight line through the origin.

Displacement Distance in a given direction.

Dissipated Describes energy that is not usefully transferred and is stored in less useful ways.

Driving force Force of a vehicle that makes it move (sometimes referred to as motive force).

Dynamo A direct-current generator.

Earth wire The wire in a mains cable used to connect the metal case of an appliance to earth.

Echo Reflection of sound that can be heard.

Efficiency Useful energy transferred by a device ÷ total energy supplied to the device.

Effort The force applied to a device used to raise a weight or move an object.

Elastic A material is elastic if it is able to regain its shape after it has been squashed or stretched.

Electric field A charged object (X) creates an electric field around itself, which causes a non-contact force on any other charged object in the field.

Electromagnet An insulated wire wrapped round an iron bar that becomes magnetic when there is a current in the wire.

Electromagnetic induction The process of inducing a potential difference in a wire by moving the wire so it cuts across the lines of force of a magnetic field.

Electromagnetic spectrum The continuous spectrum of electromagnetic waves.

Electromagnetic waves Electric and magnetic disturbances that transfer energy from one place to another.

Electrons Tiny, negatively charged particles that move around the nucleus of an atom.

Extension The increase in length of a spring (or a strip of material) from its original length.

Fleming's left-hand rule A rule that gives the direction of the force on a current-carrying wire in a magnetic field according to the directions of the current and the field.

Focal length The distance from the centre of a lens to the point where light rays parallel to the principal axis are focused (or, in the case of a diverging lens, appear to diverge from).

Force A force (in newtons, N) can change the motion of an object.

Force multiplier A lever used so that a weight or force can be moved by a smaller force.

Free-body force diagram A diagram that shows the forces acting on an object without any other objects or forces shown.

Freezing point The temperature at which a pure substance freezes.

Frequency The number of wave crests passing a fixed point every second.

Friction The force opposing the relative motion of two solid surfaces in contact.

Fuse A fuse contains a thin wire that melts and cuts the current off if too much current passes through it.

Gamma radiation Electromagnetic radiation emitted from unstable nuclei in radioactive substances.

Generator effect The production of a potential difference using a magnetic field.

Geothermal energy Energy that comes from energy released by radioactive substances deep within the Earth.

Gradient (of a straight line graph) Change of the quantity plotted on the y-axis divided by the change of the quantity plotted on the x-axis.

Gravitational field strength, g The force of gravity on an object of mass 1 kg (in newtons per kilogram, N/kg). It is also the acceleration of free fall.

Half-life Average time taken for the number of nuclei of the isotope (or mass of the isotope) in a sample to halve.

Hooke's law The extension of a spring is directly proportional to the force applied, as long as its limit of proportionality is not exceeded.

Induced magnetism Magnetisation of an unmagnetised magnetic material by placing it in a magnetic field.

Inertia The tendency of an object to stay at rest or to continue in uniform motion.

Infrared radiation Electromagnetic waves between visible light and microwaves in the electromagnetic spectrum.

Internal energy The energy of the particles of a substance due to their individual motion and positions.

Ion A charged atom or molecule.

Ionisation Any process in which atoms or molecules become charged.

Irradiated Describes an object that has been exposed to ionising radiation.

Isotopes Atoms with the same number of protons and different numbers of neutrons.

Latent heat The energy transferred to or from a substance when it changes its state.

Light-dependent resistor (LDR) A resistor whose resistance depends on the intensity of the light incident on it.

Light-emitting diode (LED) A diode that emits light when it conducts.

Limit of proportionality The limit for Hooke's law applied to the extension of a stretched spring.

Live wire The mains wire that has a voltage that alternates in voltage (between +325 V and 325 V in Europe).

Load The weight of an object raised by a device used to lift the object, or the force applied by a device when it is used to shift an object.

Longitudinal wave A wave in which the vibrations are parallel to the direction of energy transfer.

Magnetic field The space around a magnet or a current-carrying wire.

Magnetic field lines Lines in a magnetic field along which a magnetic compass points – also called lines of force.

Magnetic flux density A measure of the strength of the magnetic field defined in terms of the force on a current-carrying conductor at right angles to the field lines.

Magnification The image height ÷ the object height.

Magnifying glass A converging lens used to magnify a small object, which must be placed between the lens and its focal point.

Magnitude The size or amount of a physical quantity.

Main sequence The main sequence is the life stage of a star during which it radiates energy because of fusion of hydrogen nuclei in its core.

Mass The quantity of matter in an object – a measure of the difficulty of changing the motion of an object (in kilograms, kg).

Mass number The number of protons and neutrons in a nucleus.

Mechanical waves Vibrations that travel through a substance.

Melting point Temperature at which a pure substance melts or freezes (solidifies).

Microwaves Electromagnetic waves between infrared radiation and radio waves in the electromagnetic spectrum.

Moderator Substance in a nuclear reactor that slows down fission neutrons.

Moment The turning effect of a force, defined by the equation: moment of a force (in newton metres, Nm) = force (in newtons, N) × perpendicular distance from the pivot to the line of action of the force (in metres, m).

Momentum This equals mass (in kg) × velocity (in m/s).

Motor effect When a current is passed along a wire in a magnetic field, and the wire is not parallel to the lines of the magnetic field, a force is exerted on the wire by the magnetic field.

National Grid The network of cables and transformers used to transfer electricity from power stations to consumers (i.e., homes, shops, offices, factories, etc.).

Neutral wire The wire of a mains circuit that is earthed at the local substation so its potential is close to zero.

Neutron star The highly compressed core of a massive star that remains after a supernova explosion.

Neutrons Uncharged particles of the same mass as protons. The nucleus of an atom consists of protons and neutrons.

Newton's first law of motion If the resultant force on an object is zero, the object stays at rest if it is stationary, or it keeps moving with the same speed in the same direction.

Newton's second law of motion The acceleration of an object is proportional to the resultant force on the object, and inversely proportional to the mass of the object.

Newton's third law of motion When two objects interact with each other, they exert equal and opposite forces on each other.

Normal Straight line through a surface or boundary perpendicular to the surface or boundary.

Nuclear fission The process in which certain nuclei (uranium-235 and plutonium-239) split into two fragments, releasing energy and two or three neutrons as a result.

Nuclear fission reactor A reactor that releases energy steadily due to the fission of a suitable isotope, such as uranium-235.

Nuclear fuel A substance used in nuclear reactors that releases energy due to nuclear fission.

Nuclear fusion The process where small nuclei are forced together to fuse and form a larger nucleus.

Nucleus The tiny, positively charged centre of every atom, composed of protons and neutrons.

Ohm's law The current through a resistor at constant temperature is directly proportional to the potential difference across the resistor.

Optical fibre A thin glass fibre used to transmit light signals.

Oscillate Move to and fro about a certain position along a line.

Parallel Describes components connected in a circuit so that the potential difference is the same across each one.

Parallelogram of forces A geometrical method used to find the resultant of two forces that do not act along the same line.

Physical change A change in which no new substances are produced.

Plug A plug has an insulated case and is used to connect the cable from an appliance to a socket.

Potential difference A measure of the work done or energy transferred to the lamp by each coulomb of charge that passes through it. The unit of potential difference is the volt (V).

Power The energy transformed or transferred per second. The unit of power is the watt (W).

Pressure Force per unit cross-sectional area for a force acting on a surface at right angles to the surface. The unit of pressure is the pascal (Pa) or newton per square metre (N/m^2).

Primary waves (P-waves) Longitudinal waves that push or pull on the material that they move through as they travel through the Earth.

Principal focus The point where light rays parallel to the principal axis of a lens are focused (or, in the case of a diverging lens, appear to diverge from).

Principle of moments For an object in equilibrium, the sum of all the clockwise moments about any point = the sum of all the anti-clockwise moments about that point.

Protons Positively charged particles with an equal and opposite charge to that of an electron.

Protostar The concentration of dust clouds and gas in space that forms a star.

Radiation dose The amount of ionising radiation a person receives.

Radio waves Electromagnetic waves of wavelengths greater than 0.10 m.

Radioactive contamination The unwanted presence of materials containing radioactive atoms on other materials.

Rarefaction Stretching apart.

Reactor core The thick steel vessel used to contain fuel rods, control rods, and the moderator in a nuclear fission reactor.

Real image An image formed by a lens that can be projected on a screen.

Red giant A star that has expanded and cooled, resulting in it becoming red and much larger and cooler than it was before it expanded.

Red supergiant A star much more massive than the Sun will swell out after the main sequence stage to become a red supergiant before it collapses.

Red-shift An increase in the wavelength of electromagnetic waves emitted by a star or galaxy due to its motion away from us. The faster the speed of the star or galaxy, the greater the red-shift is.

Reflection The change of direction of a light ray or wave at a boundary when the ray or wave stays in the incident medium.

Refraction The change of direction of a light ray when it passes across a boundary between two transparent substances (including air).

Renewable Describes energy from natural sources that is always being replenished so it never runs out.

Resistance Resistance (in ohms, Ω) = potential difference (in volts, V) ÷ current (in amperes, A).

Resultant force A single force that has the same effect as all the forces acting on the object.

Scalar A physical quantity, such as mass or energy, that has magnitude only (unlike a vector, which has magnitude and direction).

Secondary waves (S-waves) Transverse waves that shake the Earth from side to side as they pass through.

Seismic waves Shock waves that travel through the Earth and across its surface as a result of an earthquake.

Series Describes components connected in a circuit in such a way that the same current passes through them.

Solenoid A long coil of wire that produces a magnetic field in and around the coil when there is a current in the coil.

Specific heat capacity Energy needed to raise the temperature of 1 kg of a substance by 1 °C.

Specific latent heat of fusion Energy needed to melt 1 kg of a substance with no change of temperature.

Specific latent heat of vaporisation Energy needed to boil away 1 kg of a substance with no change of temperature.

Specular reflection Reflection from a smooth surface. Each light ray is reflected in a single direction.

Speed The speed of an object (metres per second) = distance moved by the object (metres) ÷ time taken to move the distance travelled (seconds).

Split-ring commutator Metal contacts on the coil of a direct current motor that connects the rotating coil continuously to its electrical power supply.

Spring constant Force per unit extension of a spring.

Static electricity Electric charge stored on insulated objects.

Step-down transformer An electrical device used to step down the size of an alternating potential difference.

Step-up transformer An electrical device used to step up the size of an alternating potential difference.

Stopping distance The distance travelled by a vehicle in the time it takes for the driver to think and brake.

Supernova The explosion of a massive star after fusion in its core ceases and the matter surrounding its core collapses onto the core and rebounds.

Tangent A straight line drawn to touch a point on a curve so it has the same gradient as the curve at that point.

Terminal velocity The velocity reached by an object when the drag force on it is equal and opposite to the force making it move.

Thermal conductivity A property of a material that determines the energy transfer through it by conduction.

Thermistor A resistor whose resistance depends on its temperature.

Thinking distance The distance travelled by the vehicle in the time it takes the driver to react.

Three-pin plug A three-pin plug has a live pin, a neutral pin, and an earth pin.

Transformer An electrical device used to change an (alternating) voltage. See also **Step-up transformer** and **Step-down transformer**.

Transmitted Describes a wave passing through a substance.

Transverse wave A wave where the vibration is perpendicular to the direction of energy transfer.

Ultrasound wave Sound wave at frequency greater than 20 000 Hz (the upper frequency limit of the human ear).

Ultraviolet radiation Electromagnetic waves between visible light and X-rays in the electromagnetic spectrum.

Upthrust The upward force that acts on a body partly or completely submerged in a fluid.

Useful energy Energy transferred to where it is wanted in the way that is wanted.

Vector A physical quantity, such as displacement or velocity, that has a magnitude and a direction (unlike a scalar, which has magnitude only).

Velocity Speed in a given direction (in metres/second, m/s).

Vibrate Oscillate (move to and fro) rapidly about a certain position.

Virtual image An image, seen in a lens or a mirror, from which light rays appear to come after being refracted by a lens or reflected by a mirror.

Wasted energy Energy that is not usefully transferred.

Wave speed The distance travelled per second by a wave crest or trough.

Wavelength The distance from one wave crest to the next.

Weight The force of gravity on an object (in newtons, N).

White dwarf A star that has collapsed from the red giant stage to become much hotter and denser.

White light Light that includes all the colours of the spectrum.

Work The energy transferred by a force. Work done (joules, J) = force (newtons, N) × distance moved in the direction of the force (metres, m).

X-rays Electromagnetic waves shorter in wavelength than ultraviolet radiation, produced by X-ray tubes.

Answers

P1.1 Changes in energy stores

1 elastic potential energy
2 gravitational potential energy

P1.2 Conservation of energy

1 an object or a group of objects in which no energy transfers take place out of or into the energy stores of the system
2 energy is transferred from the chemical energy store of the battery (by the electric current in the wire between the battery and bulb) to the thermal energy store of the surroundings (by light waves and by heating)

P1.3 Energy and work

1 joule, J
2 7200 J (or 7200 N m)

P1.4 Gravitational energy stores

1 it decreases
2 newtons per kilogram, N/kg

P1.5 Kinetic energy and elastic energy stores

1 3 J
2 500 J

P1.6 Energy dissipation

1 it is dissipated to the surroundings, which will warm up
2 energy cannot be created or destroyed, so it cannot be lost; but is wasted when it is not usefully transferred

P1.7 Energy and efficiency

1 0.4 or 40%
2 it increases

P1.8 Electrical appliances

1 by heating the surroundings
2 **a** by heating the bread
b by heating the surroundings and the body of the toaster

P1.9 Energy and power

1 2200 W
2 300 W

P1 Summary questions

1 elastic potential energy [1]
2 decrease in the chemical energy store of the child's muscles; [1] increase in the gravitational potential energy store of the ball [1]
3 90 000 J [1]
4 change in gravitational potential energy = 0.45 × 9.8 × 3.0 [1]
= 13 J (2 s.f.) [1]
5 chemical energy store in student's muscles decreases as they stretch the spring; [1] elastic potential energy store of the spring increases as the spring stretches; [1] as the spring is released the elastic potential energy store of the spring decreases and the thermal energy store of the surroundings increases [1]
6 work is only done when a force moves an object through a distance; [1] because the car does not move, no work is done on it [1]
7 50 g = 0.05 kg and 8.0 cm/s = 0.08 m/s [1]
$\text{kinetic energy} = \frac{1}{2} \times 0.05 \times (0.08)^2$ [1]
$= 1.6 \times 10^{-4}$ J [1]
8 $\text{power} = \frac{200\,\text{N} \times 12\,\text{m}}{15\,\text{s}}$ [1]
= 160 W [1]
9 when the rope is slack, energy is transferred from the jumper's gravitational potential energy store to their kinetic energy store; [1] as the rope tightens, the jumper's kinetic energy store decreases [1]; and the rope's elastic potential energy store increases; [1] when the jumper comes to a stop, all the energy has all been transferred to the elastic potential energy store of the rope [1]
10 work done on the trolley = weight of the trolley × increase in height; [1]
work done by the person = force × distance moved along the slope; [1]
some of the work done by the person is wasted by doing work against friction [1]
11 total energy input = 42 J + 28 J
= 70 J [1]
$\text{motor efficiency} = \frac{28}{70} \times 100$ [1]
= 40% [1]
12 $k = \frac{2E_p}{e^2}$ [1]
$= \frac{2 \times 1.2}{e^2}$ [1]
= 375 N/m [1]

P2.1 Energy transfer

1 metals are good conductors so will help the food to heat up; wood is a good insulator so will protect you from getting burnt
2 air is a poor thermal conductor

P2.2 Infrared radiation

1 a type of electromagnetic wave
2 the higher the object's temperature, the greater the rate at which it emits infrared radiation
3 the intensity of an object's black body radiation has a peak at a certain wavelength; this peak wavelength depends on the object's temperature

P2.3 More about infrared radiation

1 matt black surfaces
2 it may be reflected back into space, absorbed by the earth's atmosphere, or absorbed by the earth's surface

P2.4 Specific heat capacity

1 8400 J
2 490 J

P2.5 Heating and insulating buildings

1 black surfaces are good absorbers of infrared radiation
2 to ensure maximum exposure to infrared radiation

P2 Summary questions

1 copper [1]
2 matt black surfaces [1]
3 the temperature difference across the material; [1] the thickness of the material; [1] the thermal conductivity of the material [1]
4 silver foil will reflect infrared radiation from the radiator back into the room, [1] so the infrared radiation will not be lost through the walls of the house, [1] which saves energy and reduces fuel bills [1]
5 thermal conductors allow energy to be transferred through them easily; [1] thermal insulators do not allow energy to be transferred through them easily [1]
6 $E = 2.5 \times 490 \times 3$ [1]
= 3675 J [1]
7 the starting temperature of the water in each can; [1] the thickness of each insulating material wrapped around each can; [1] the volume of water in each can [1]
8 the greater the specific heat capacity of the blocks, the more energy is required to increase their temperature by 1°C, [1] which means that more energy is stored in the thermal energy store of the blocks, [1] and so more energy is transferred to the room during the day
9 an object that absorbs all the infrared radiation which is incident on it; [1] and is also the best possible emitter of infrared radiation [1]
10 the rate at which radiation from the Sun is reflected or absorbed, since a higher rate of absorption will increase the Earth's temperature; [1] the rate at which the radiation is emitted from the Earth's surface, since a lower rate of emitted radiation will increase the Earth's temperature; [1] the rate at which radiation is emitted from the Earth's atmosphere into space, as a lower rate of emitted radiation will increase the Earth's temperature [1]
11 when the filament is at a constant temperature, it emits radiation across a continuous range of wavelengths, which are seen as different colours; [1] but the intensity of the radiation it emits has a peak at a certain wavelength, which depends on the filament's temperature; [1] as the potential difference across the filament increases, its temperature increases and so the wavelength (and colour) of the peak radiation changes [1]
12 $c = \frac{E}{m\Delta\theta}$ [1]
$= \frac{16200}{1.5 \times (32 - 20)}$ [1]
= 900 J/kg °C [1]

P3.1 Energy demands

1 an energy source that is being used up faster than it is being replaced
2 coal, oil, gas
3 86%

P3.2 Energy from wind and water

1 gravitational potential energy
2 the wind does not always blow; and the speed and direction of the wind vary

P3.3 Power from the Sun and the Earth

1 single solar cells do not produce much power
2 there are only a few places in the world where the hot rocks are close enough to the surface to be accessible

P3.4 Energy and the environment

1 a large, exposed area, which is away from buildings and people
2 waves are dependent on the weather, and so are not reliable
3 78%

P3.5 Big energy issues

1 since it is colder in winter, people will have their heating on more, and use more hot water; the winter daylight hours are shorter so people need to keep lights on for longer
2 during the day, people are using more household appliances such as kettles, computers, and lights
3 least at 4 a.m.; greatest at 8 p.m.

P3 Summary questions

1 one that is replaced as quickly as it is used up [1]
2 an electricity generator, driven by the wind, on top of a tall tower [1]
3 nuclear, [1] coal-fired [1]
4 water is collected in a reservoir at the top of a hill; [1] when the water is allowed to flow downhill it turns turbines at the bottom of the hill which are connected to electricity generators [1]
5 the Sun does not shine at night; [1] and in many countries the amount of sunlight received at the ground varies with the seasons and weather [1]
6 waste vegetable oil; [1] methane; [1] woodchip [1]
7 burning coal releases carbon dioxide, [1] which is a greenhouse gas, [1] and also releases sulphur dioxide, [1] which causes acid rain [1]
8 there must be a significant difference in the water level between high and low tides, [1] such as a broad estuary that rapidly becomes narrower up-river, [1] and there must be space to build a barrage across the river [1]
9 water is pumped down to hot rocks under the surface of the Earth; [1] the water is heated by the rocks [1] to produce steam, [1] which drives turbines connected to electricity generators at ground level [1]
10 conditions will generally be right for either one or the other to produce electricity; [1] on sunny days, the solar panel will produce electricity to power the sign; [1] on cloudy days (which are generally windy) and at night, the wind turbine can produce electricity to power the sign [1]
11 2 MW = 2000 kW [1]
energy = power × time
= 2000 kW × 24 hours [1]
= 48 000 kW h [1]
12 when electricity demand is low, electricity from other power stations is used at a pumped storage scheme to pump water uphill into the reservoir; [1] when there is a sudden surge in demand for electricity, the water can be allowed to run downhill to generate electricity, [1] with little start-up time required [1]

Section 1 Practice questions

01.1 conduction [1]
01.2 heating the kettle itself; [1] heating the surroundings [1]
01.3 $E = 0.300 \times 4200 \times 75$ [1]
$= 94\,500\ \text{J}$ [2]
02.1 gravitational potential; [1] elastic potential; [1] kinetic [1]
02.2 $\frac{1}{2} \times 24 \times 0.025^2$ [1]
$= 0.0075\ \text{J}$ [1]
02.3 spring might snap [1] and hit face; [1] wear goggles to protect eyes [1]
03.1 30% [1]
03.2 58 000 J [1]
03.3 reducing external noise [1]
03.4 both the plastic and the air are poor conductors [1] so thermal conductivity of the cavity is reduced; [1] energy transfer through walls is reduced [1] so rate of cooling of the house is reduced [1]
04.1 two from: wind, waves, tides, falling water, geothermal [2]
04.2 to change the direction of the mirrors as the Sun moves across the sky, [1] so that the tower is always receiving maximum sunlight [1]
04.3 advantages – two from: solar power station is non-polluting, coal-fired power station is polluting; sunlight is free, whereas coal costs money to extract and mining can cause damage to environment; supplies of coal will run out, whereas sunlight is a renewable source of energy [2]
disadvantages – two from: solar power stations only run during the day, but coal-fired power stations generate electricity all the time; solar power stations only work well in places with a lot of sunlight, whereas coal-fired power stations can be used wherever the coal can be transported; a coal-fired power station can generate much more power than a solar power station [2]
05.1 electrical energy to kinetic energy (of the fan and the air); [1] kinetic energy to thermal energy (heating the air and your hair) [1]
05.2 energy transferred by sound waves; [1] heating the body of the hairdryer [1]
05.3 $\text{efficiency} = \frac{\text{useful power output}}{\text{total power input}}$ [1]

$0.85 = \frac{\text{useful power output}}{1600}$

useful power output = 1360 (W) [1]
useful energy output = 1360 × 3 × 60 [1]
= 244 800 (J) [1]
06.1 material box is made from [1]
06.2 time taken for thermometer reading to increase by 5 °C [1]
06.3 two from: same number of drinks cans in each box; same initial temperature in each box; same external temperature [2]
06.4 slowest energy transfer is through the expanded polystyrene; [1] so it is the best insulator [1]
06.5 on a sunny beach the expanded polystyrene box will absorb infrared radiation; [1] foil will prevent this because infrared radiation is reflected by shiny surfaces [1]

P4.1 Electrical charges and fields

1 by gaining electrons
2 they repel each other

P4.2 Current and charge

1

2 2 A

P4.3 Potential difference and resistance

1 voltmeter
2 ohm, Ω

P4.4 Component characteristics

1 the resistance increases
2 when the p.d. becomes negative, the current becomes negative too; so the current–potential difference graph is reflected in the x-axis (that is, the axis which shows p.d.)

P4.5 Series circuits

1 all the components connected in series stop working too
2 add the resistances of the individual components in the circuit

P4.6 Parallel circuits

1 the other components in the circuit continue to work
2 the p.d. across every parallel branch is equal to the supply p.d.

P4 Summary questions

1 negative [1]
2 33 Ω [1]
3 $I = \frac{18}{4}$ [1]
$= 4.5\ \text{A}$ [1]
4 $R = \frac{24}{4.0}$ [1]
$= 6.0\ \Omega$ [1]
5 when two insulators are rubbed together, electrons are transferred from one insulator to the other; [1] the insulator that loses electrons is left with fewer electrons than protons, [1] giving it an overall positive charge [1]
6 the ammeter should be connected in series with the resistor; [1] the voltmeter should be connected in parallel with the resistor [1]
7 as the resistance of the variable resistor increases, the variable resistor becomes a greater proportion of the total resistance in the circuit; [1] since the current is the same throughout the circuit, the p.d. across the variable resistor increases

8

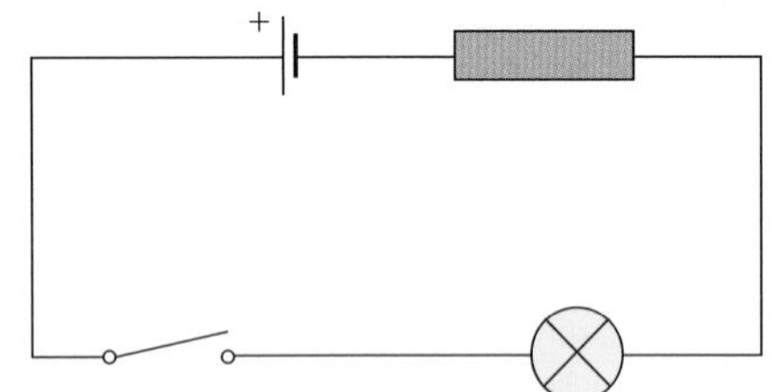

[all 4 correct circuit symbols – 2 marks; 3 correct – 1 mark; 1 mark for correctly showing a series connection]

9

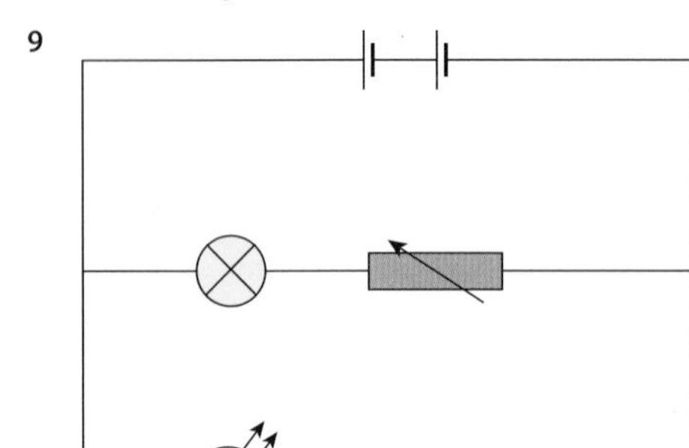

bulb and variable resistor shown with correct circuit symbols [1] connected in series; [1] LED and resistor shown with correct circuit symbols [1] connected in series; [1] battery shown with correct circuit symbol and connected in parallel with series branches [1]

10 an ohmic conductor obeys Ohm's law, so the current through it is directly proportional to the p.d. across it, [1] provided its temperature is constant [1]

11 $E = 4.5 \times 8.0$ [1]

$= 36\,J$ [1]

12

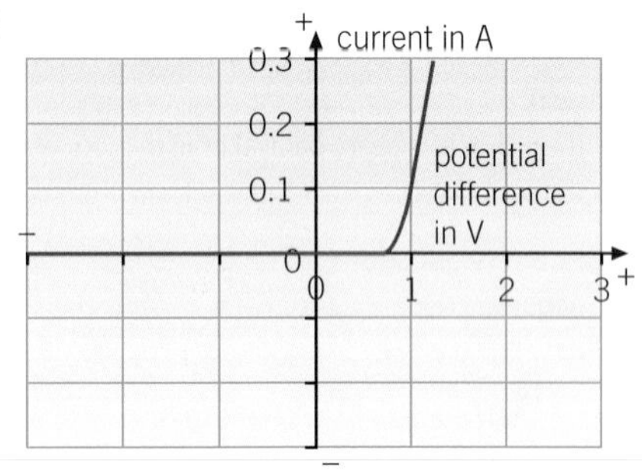

in the forward direction, the line curves towards the *y*-axis as the current is not directly proportional to the p.d.; [1] in the reverse direction, the current is virtually zero as the diode's resistance is extremely high [1]

P5.1 Alternating current

1 current that passes around a circuit in one direction only

2 0V

P5.2 Cables and plugs

1 plastic is a good insulator and is flexible so that the wires can be bent

2 the earth pin, so that it connects to a socket first

P5.3 Electrical power and potential difference

1 a thin wire that will melt if the current passing through it exceeds a given value

2 $P = \frac{E}{T}$

$= \frac{90000}{30}$

$= 3000\,W = 3\,kW$

P5.4 Electrical currents and energy transfer

1 $Q = It$

$= 2.0 \times 5 \times 60$

$= 600\,C$

2 $E = VQ$

$= 230 \times 200$

$= 46\,000\,J$

P5.5 Appliances and efficiency

1 $\frac{5}{25} = 0.2$ or 20%

2 $E = P \times t$

$= 2000 \times 3 \times 60$

$= 360000$

P5 Summary questions

1 50 Hz [1]

2 **a** brown [1] **b** blue [1] **c** green/yellow [1]

3 plastic is an insulator [1]

4 $P = VI$

$= 230 \times 10$ [1]

$= 2300\,W$ [1]

5 a current which continually changes direction [1] from a positive potential difference to a negative potential difference [1]

6 $E = VQ$

$= 230 \times 350$ [1]

$= 80\,500\,J$ [1]

7 the outer case of a hairdryer is made from plastic, which is an insulator; [1] so the cable does not need an earth wire [1]

8 brass is a good conductor; [1] it is hard so the pins will not bend or be easily damaged; [1] and it does not rust or oxidise [1]

9 $t = \frac{1}{f}$

$= \frac{1}{60}$ [1]

$= 0.017\,s$ [1]

10 $I = \frac{P}{V}$

$= \frac{650}{230}$

$= 2.83\,A$ [1]

a fuse with a rating just higher than the normal working current should be used; [1] so the 3 A fuse is suitable [1]

11 useful output power = efficiency × total input power

$= 0.83 \times 1800$ [1]

$= 1500\,W$ (2 s.f.) [1]

12

[1 mark for correct shape of curve; 1 mark for correct peak values; 1 mark for crossing time axis in correct places]

P6.1 Density

1 g/cm^3

2 $2700\,kg/m^3$

P6.2 States of matter

1 the particles in a solid vibrate about fixed positions so that the solid has a fixed shape; the particles in a liquid can move about at random, so a liquid doesn't have a fixed shape and can flow

2 the particles in a liquid are in contact with each other but can move about at random; the particles in a gas are usually far apart and move about at random – much faster than the particles in a liquid

P6.3 Changes of state

1 it changes state from a solid to a liquid, or from a liquid to a solid

2 the energy transferred to or from a substance as it changes state

P6.4 Internal energy

1 it decreases

2 solid state

P6.5 Specific latent heat

1 the energy required to change 1 kg of a substance from a liquid to a vapour, with no change in temperature

2 J/kg

P6.6 Gas pressure and temperature

1 it decreases

2 it increases

P6.7 Gas pressure and volume

1 pascal, Pa

2 it decreases

P6 Summary questions

1 the mass of the substance per unit volume [1]

2 gas [1]

3 the freezing point of a substance is the temperature at which it changes from a liquid state to a solid state, or vice versa; [1] the boiling point of a substance is the temperature at which it changes from a liquid state to a gas state, or vice versa [1]

4 $L_f = \frac{E}{m}$

$= \frac{8350}{0.25}$ [1]

$= 33\,400\,J/kg$ [1]

5 $\rho = \frac{m}{V}$

$= \frac{63.2}{2^3}$ [1]

$= 7.9\,g/cm^3$ [1]

6 the particles in a solid vibrate about fixed positions; [1] so the solid has a fixed shape; [1] the particles in a gas are usually far apart; [1] and move about very quickly at random [1]

7 the particles in a gas move at high speeds in random directions; [1] so they collide with each other and with the walls of their container; [1] during these collisions, the particles exert a force, and hence a pressure [1] on the walls of the container [1]

8

gas pressure in Pa

0

0

temperature in °C

[straight line – 1; positive gradient – 1; non-zero intercept on y-axis – 1]

9 $V = \frac{m}{\rho}$

$= \frac{60}{2400}$

$= 0.025\,m^3$ [1]

thickness $= \frac{\text{volume}}{\text{area}}$

$= \frac{0.025}{0.5}$ [1]

$= 0.05\,m$ [1]

10 boiling takes place throughout a liquid [1] when the temperature of the liquid is at its boiling point; [1] evaporation takes place from the surface of a liquid [1] when the temperature of the liquid is below its boiling point [1]

11 the total energy stored [1] in the kinetic energy store and the potential energy store [1] of the particles in the substance [1]

12 **a** the pressure increases (since pV = constant); [1] the temperature of the gas does not change (since work done on the gas is transferred, by heating, to the surroundings) [1]

b the pressure increases (since pV = constant); [1] the temperature increases (since the work done on the gas increases both the internal energy of the gas and its temperature) [1]

P7.1 Atoms and radiation

1 alpha, beta, and gamma

2 it stays the same

P7.2 The discovery of the nucleus

1 because an atom consists mostly of empty space

2 in the nucleus

P7.3 Changes in the nucleus

1 +2

2 −1

P7.4 More about alpha, beta, and gamma radiation

1 alpha

2 gamma

P7.5 Activity and half-life

1 it decreases

2 it has decreased to a quarter of its original value

P7.6 Nuclear radiation in medicine

1 alpha radiation cannot be monitored with a detector outside the body; it is also very ionising, so can cause damage within the body

2 the half-life is long enough to complete the imaging, but short enough to avoid exposing the patient to unnecessary radiation

P7.7 Nuclear fission

1 uranium-238, with 2–3% uranium-235

2 fission that takes place when a neutron is absorbed by a uranum-235 nucleus or a plutonium-239 nucleus

P7.8 Nuclear fusion

1 nuclear fusion

2 using a magnetic field

P7.9 Nuclear issues

1 radon gas and cosmic rays

2 it emits radiation, which may cause cancer and have other harmful effects

P7 Summary questions

1 it does not change [1]

2 empty space [1]

3 atomic number is the number of protons in the nucleus; [1] mass number is the number of protons and neutrons in the nucleus [1]

4 the process where two smaller nuclei join together ('fuse') [1] to make a larger nucleus [1]

5 94 protons; [1] 145 neutrons [1]

6 alpha particles are the least penetrating – they can travel only a few centimetres through air [1] and are stopped by a thin sheet of paper; [1] beta particles can travel a few metres through air [1] and are stopped by a thin sheet of aluminium; [1] gamma rays are the most penetrating – they have an unlimited range in air [1] and several centimetres of lead or several metres of concrete are needed to absorb most of the gamma radiation [1]

7 32 days = 4 half-lives [1]

mass remaining after 32 days $= \frac{24}{2^4}$ [1]

$= 1.5\,mg$ [1]

8

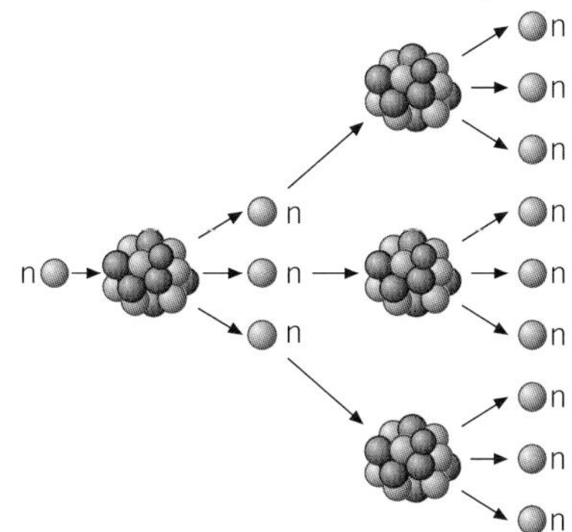

[initial neutron and uranium nucleus – 1; further neutrons emerging – 1; those neutrons meeting further uranium nuclei and producing more neutrons – 1]

9 45 hours = 3 half-lives [1]

number of unstable nuclei after 45 hours

$= \frac{800\,000}{2^3}$ [1]

$= 100\,000$ [1]

10 the isotope emits gamma radiation, which penetrates deep into the body to reach a tumour and destroy the cancer cells [1]; its half-life of 5 years means that the source will continue to produce a sufficient dose of radiation to kill all the cancer cells [1] but will not expose the patient to high levels of radiation for years to come [1]

11 induced fission occurs when a neutron is absorbed by a uranum-235 nucleus or a plutonium-239 nucleus [1] and the nucleus splits; [1] spontaneous fission occurs without a neutron being absorbed [1]

12 fast-moving neutrons don't go on to cause further fission [1] so a moderator (usually water) is used to slow the high-speed neutrons down; [1] control rods in the core of a nuclear reactor absorb surplus neutrons [1] to prevent uncontrolled fission [1]

Section 2 Practice questions

01.1 $V = IR$ [1]

$= 0.5 \times 2$ [1]

$= 1\,(V)$ [1]

01.2

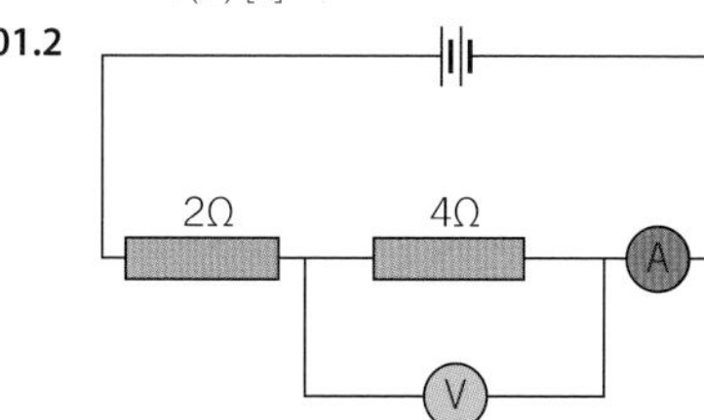

correct voltmeter symbol [1] in parallel across 4Ω resistor [1]

01.3 total p.d. = 1 + 2 = 3V [1]

02.1 alpha [1]

02.2 gamma [1]

02.3 alpha [1]

02.4 92 protons; [1] 146 neutrons [1]

03 measure out a known volume of the liquid in a measuring cylinder; [1] measure the mass of an empty beaker using a balance; [1] pour the known volume of liquid into the beaker and measure the mass of beaker plus liquid; [1] subtract the mass of the empty beaker to give the mass of the liquid; [1] calculate the density of the liquid using density $= \frac{\text{mass}}{\text{volume}}$ [1]

04.1 $P = VI$ [1]

$= 50 \times 4.2$ [1]

$= 210$ [1] W (or watts) [1]

04.2 100 g = 0.1 kg [1]

$L_F = \frac{6300}{0.1}$ [1]

$= 63\,000$ (J/kg) [1]

05.1 the gas molecules continually collide with the walls of the piston and bounce off; [1] each collision causes a force on the walls; [1] pressure $= \frac{\text{force}}{\text{area}}$ so the millions of collisions each second cause a pressure on the walls [1]

05.2 $101 \times 1200 = 303 \times V_2$ [1]

$V_2 = 400\,cm^3$ [1]

05.3 when the force is applied to push the piston in, work is done; [1] if the compression is done slowly, there is sufficient time for the energy to be transferred to the surroundings by heating; [1] so the internal energy store and temperature do not increase [1]

06.1 total irradiation time = 283 s × 120

$= 33\,960$ s [1]

$= \frac{33\,960}{60 \times 60}$ hours [1]

$= 9.43$ hours [1]

06.2 total irradiation time increases as relative density increases; [1] the denser the instrument the more difficult for gamma radiation to penetrate it, [1] so the greater total irradiation time needed for sterilisation [1]

06.3 so the medical instrument stays sterile [1] until the bag is opened [1]

06.4 gamma radiation is dangerous to people if it passes outside the chamber, so must be absorbed by the chamber walls; [1] gamma radiation is very penetrating, so the walls must be made from thick concrete in order to absorb most of it [1]

06.5 gamma radiation kills microorganisms on the food [1] without damaging the food, [1] extending its shelf life and reducing food waste [1]; but some people worry that the food is harmful once irradiated; [1] and should have the choice whether or not they want to consume irradiated food; [1] and so may not purchase it [1]

P8.1 Vectors and scalars

1 a scalar is a quantity with magnitude only; a vector is a quantity with both magnitude and direction

2 7 cm

P8.2 Forces between objects

1 newton, N

2 vertically downwards

P8.3 Resultant forces

1 10 N in the same direction as the two forces

2 4 N in the same direction as the 7 N force

P8.4 Moments at work

1 12 N m

2 40 N

3 apply a greater force; and/or apply the force closer to the end of the handle

P8.5 More about levers and gears

1 to change the turning effect of a force

2 the smaller gear wheel experiences a smaller turning moment and a greater turning speed than the larger gear wheel

P8.6 Centre of mass

1 at the point where its axes of symmetry meet

2 to increase their stability

P8.7 Moments and equilibrium

1 the weight of the person acts downwards through the pivot

2 move away from the centre of the seesaw

P8.8 The parallelogram of forces

1 1 cm represents 3 N

2 6.5 N at 26° to the 3 N force

P8.9 Resolution of forces

1 parts of a force that, when acting together, have the same effect as the original force

2 horizontal component = 12 N
vertical component = 5 N

P8 Summary questions

1 e.g. scalar – mass; [1] vector – force [1]

2 a force which acts on objects only when the objects touch each other [1]

3 moment = 35 × 2.1 [1]
= 73.5 N m [1]

4 at the 50 cm mark; [1] halfway across the width of the rule [1]

5 a quantity that has both magnitude and an associated direction [1]

6 when two objects interact with each other they exert equal [1] and opposite [1] forces on each other

7 the object is either stationary [1] or moving at a constant speed [1] in a straight line [1]

8 Adam's moment = 400 × 2.5 = 1000 N m
Ben's moment = 450 × 2.0 = 900 N m
Adam's turning moment about the pivot is greater than Ben's, so Adam will move downwards

9 scalar quantities: distance, speed, time, mass [2 marks; 2 correct – 1 mark]
vector quantities: displacement, velocity, weight, acceleration
[2 marks; 2 correct – 1 mark]

10 when a car is being driven forwards, each tyre exerts a force on the ground which pushes backwards; [1] Newton's third law [1] states that the ground exerts an equal and opposite force on each tyre, which pushes the car forwards; [1] the force between each tyre and the ground is caused by friction between the tyre and the ground at the point of contact [1]

11 force $= \frac{\text{moment}}{\text{distance from pivot}}$ [1]

$= \frac{4.2}{0.35}$ [1]

= 12 N [1]

12

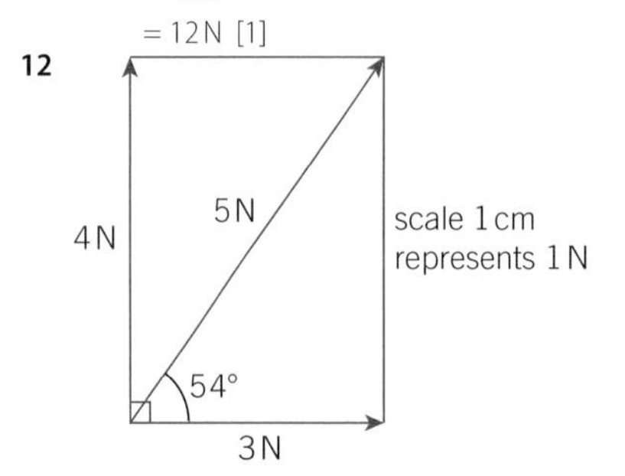

[3 N force and 4 N force drawn to scale – 1; angle of 90° between them – 1; completed parallelogram and resultant – 1; resultant of correct size and at correct angle (5 N at 54° to 3 N force) – 1]

P9.1 Speed and distance–time graphs

1 8 m/s

2 speed

P9.2 Velocity and acceleration

1 speed is a scalar quantity and has magnitude only; velocity is a vector quantity – it is defined as speed in a specified direction

2 metres per second squared, m/s^2

P9.3 More about velocity–time graphs

1 constant velocity

2 moves with constant velocity for 10 seconds; then decelerates at a constant rate during the next 5 seconds until it comes to a stop

P9.4 Analysing motion graphs

1 speed

2 increasing speed

3 increasing acceleration

P9 Summary questions

1 $v = \frac{s}{t}$

$= \frac{1170}{90}$ [1]

= 13 m/s [1]

2 $a = \frac{\Delta v}{t}$

$= \frac{20}{16}$ [1]

= 1.25 m/s^2 [1]

3 acceleration at a constant rate; [1] then constant speed [1]

4 acceleration [1]

5 $v = \frac{s}{t}$

$= \frac{2400}{60}$ [1]

= 40 m/s [1]

6 $a = \frac{\Delta v}{t}$

$= \frac{25 - 15}{5.0}$ [1]

= 2.0 m/s^2 [1]

7 straight horizontal line; [1] then straight line sloping downwards [1] to zero [1]

8 $a = \frac{\Delta v}{t}$

$= \frac{12 - 4}{10}$ [2]

= 0.80 m/s^2 [1]

9 $\Delta v = a\,t$

= 1.5 × 8.0 [1]

= 12.0 [1]

initial speed = 0 m/s; so final speed = 12.0 m/s [1]

10 108 km/h = 108 000 m/h

$= \frac{108\,000}{60 \times 60}$ m/s

= 30 m/s

11 distance travelled = area under graph

$= (4 \times 10) + (\frac{1}{2} \times 10 \times 8)$

= 80 m

12 tangent to curve at 15 s drawn correctly [1]
speed at 15 s = gradient of tangent to curve at 15 s [1]

$= \frac{185 - 0}{20 - 7.5}$ [1]

= 14.8 m/s [1]

P10.1 Forces and acceleration

1 it increases
2 2400 N

P10.2 Weight and terminal velocity

1 the only force which initially acts on the object is its weight, so there is a resultant downwards force
2 it decreases to zero

P10.3 Forces and braking

1 zero
2 stopping distance = thinking distance + braking distance

P10.4 Momentum

1 zero
2 $1500 \times 20 = 30\,000$ kg m/s
3 in a closed system, the total momentum before an event is the same as the total momentum after the event

P10.5 Using conservation of momentum

1 zero
2 both students have momentum of equal magnitude and opposite direction
3 36 000 kg m/s

P10.6 Impact forces

1 crumple in a collision, and so increase the impact time of the collision
2 the change in momentum; and the impact time

P10.7 Safety first

1 they would continue to travel with the same velocity as before the collision and so hit the dashboard or windscreen
2 as the passenger continued to travel forwards, the narrow seat belt would apply a large force over a small area, which would probably cause injury

P10.8 Forces and elasticity

1 one that does not regain its shape after it has been deformed
2 its graph of applied force against extension begins to curve

P10 Summary questions

1 $F = ma$
$= 45 \times 11$ [1]
$= 495$ N [1]
2 $W = mg$
$= 55 \times 9.8$ [1]
$= 539$ N [1]
3 worn tyres; [1] worn brakes [1]
4 a straight line; [1] through the origin [1]
5 $p = mv$
$= 0.18 \times 4$ [1]
$= 0.72$ kg m/s [1]
6 [2]

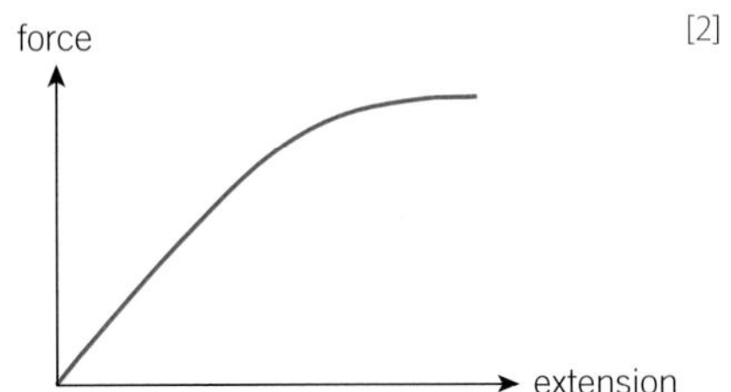

7 $F = ma = m \times \frac{\Delta v}{t}$ [1]
$= 900 \times \frac{20 - 0}{4.0}$ [1]
$= 4500$ N [1]
8 the helmet is made of a material designed to crumple on impact, [1] which increases the length of the collision time between the rider's head and the ground, [1] which decreases the impact force [1]
9 the object's tendency to remain at rest or in uniform motion [1]
10 total momentum before the bullet is fired = total momentum after the bullet is fired
$0 = (0.010 \times 860) + (3.8 \times v)$ [1]
$= 8.6 + 3.8v$
$v = \frac{-8.6}{3.8}$ [1]
$= -2.3$ m/s (that is, 2.3 m/s in opposite direction to the bullet) [1]
11 $\Delta v = \frac{F\Delta t}{m}$ [1]
$= \frac{32\,000 \times 0.75}{900}$ [1]
$= 27$ m/s [1]
car decelerates to rest, so initial speed is 27 m/s [1]
12 the soft crash mat deforms as the gymnast lands on it, [1] which increases the impact time of the collision between the gymnast and the mat; [1] and because $F = \frac{m\Delta v}{t}$, increasing the impact time decreases the impact force, [1] helping the gymnast to avoid injury [1]

P11.1 Pressure and surfaces

1 7500 Pa
2 although the weight of the person remains the same, stiletto heels substantially decrease their area in contact with the floor, thereby greatly increasing the pressure they exert on the floor

P11.2 Pressure in a liquid at rest

1 4900 Pa
2 N/m^2

P 11.3 Atmospheric pressure

1 pascal, Pa
2 it increases

P11.4 Upthrust and flotation

1 pressure increases as depth increases
2 upthrust

P11 Summary questions

1 $p = \frac{F}{A}$
$= \frac{150}{0.25}$ [1]
$= 600$ Pa [1]
2 atmospheric pressure decreases with height above sea level [1]
3 atmospheric pressure is due to air molecules colliding with surfaces and exerting a force on the surface; [1] as height above sea level increases, the air is less dense and so there are fewer molecules to collide with surfaces; [1] so the atmospheric pressure decreases [1]
4 the suction cap is pushed onto a surface, which squeezes out the air on the inside of the suction cap; [1] atmospheric pressure acts on the outside of the cap, but not on the inside [1] and pushes the cap against the surface [1]
5 $p = \frac{F}{A}$
$= \frac{65 \times 9.8}{0.020}$ [1 for multiplying m by 9.8, 1 for all correct]
$= 32\,000$ Pa [1]
6 $p = h\rho g$ [1]
$= 0.75 \times 1000 \times 9.8$ [1]
$= 7400$ Pa [1]
7 $p = h\rho g$ [1]
$= 0.75 \times 14\,000 \times 9.8$ [1]
$= 103\,000$ Pa [1]
8 take an empty plastic bottle and make several pin-sized holes around its circumference, all at the same depth; [1] fill the bottle with water and the water jet from each hole will travel the same distance from the bottle, [1] showing that the pressure of each jet is the same [1]
9 $A = \frac{F}{p}$ [1]
$= \frac{24000}{8000}$ [1]
$= 3$ m^2 [1]
10 $h = \frac{p}{\rho g}$ [1]
$= \frac{4.2 \times 10^6}{1050 \times 9.8}$ [1 for numerator, 1 for denominator]
$= 408$ m [1]
11 pressure increases with depth in a fluid; [1] so the pressure acting on an object immersed in the fluid is greater at the bottom of the object than at the top of the object; [1] so the upward force of the water on the bottom of the object is bigger than the downward force of the water on the top of the object; [1] the resultant of these two forces is the upthrust acting on the object [1]
12 the newton-meter reading goes down [1] because there is an upthrust on the object in when it is immersed in the water; [1] the magnitude of the upthrust is equal to the difference in the two newton-meter readings [1]

Section 3 Practice questions

01.1 $W = mg$ [1]
$= 1.3 \times 9.8$ [1]
$= 12.7$ N [1]
01.2 newton [1]
01.3 [1]
02.1 the gradient of a distance–time graph gives the speed; [1] the gradient of the graph is steepest between A and B [1]
02.2 80 – 40 [1]
= 40 s [1]

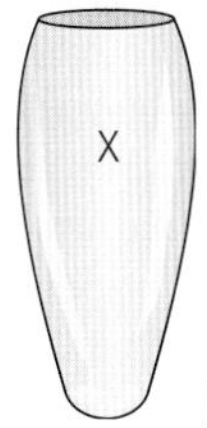

02.3 travelling at a constant speed [1]

02.4 draw a straight line [1] passing through both (0,0) and (160,400) [1]

03.1 mark the start and end points clearly on the road; [1] measure the distance between start and end points using a trundle wheel; [1] one student should stand next to the start point and raise a flag immediately as the car passes; [1] a second student should stand next to the finish point and start the stopwatch when the first student raises the flag; [1] the second student should stop the stopwatch when the car passes the finish point [1]

03.2 the students should stand next to the road, not on the road; [1] the students should wear high-visibility clothing; [1] the students should carry out the investigation on a day with good visibility [1]

03.3 speed $= \frac{\text{distance travelled}}{\text{time taken}}$ [1]

03.4 on a Wednesday morning, the number of cars breaking the speed limit drops between 8:01 and 9:30 and then increases again afterwards; [1] between 8:01 and 9:30 on a Wednesday there are likely to be many students on a road that is close to the school, so drivers will travel more slowly; [1] on a Sunday, there is no school so the number of cars breaking the speed limit will not be affected by students crossing the road; [1] the number of cars breaking the speed limit on Sunday is likely to increase gradually throughout the morning as people tend to stay indoors until later on Sundays [1]

04.1 the mass suspended from the spring [1]

04.2 weight = mass × gravitational field strength [1]

04.3 gradient of linear part of graph gives spring constant [1]

gradient $= \frac{1.0 - 0}{0.015 - 0}$ [1]

$= 66.7$ N/m [1]

04.4 approximately 12 mm; [1] up to this point the graph is a straight line [1]

05.1 $m = \frac{W}{g}$ [1]

$= \frac{686}{9.8}$ [1]

$= 70$ kg [1]

05.2 air resistance [1]

05.3 initially the skydiver is stationary so X is 0, and the only force acting on the skydiver is Y, so she accelerates downwards; [1] gradient of velocity–time graph represents acceleration, so initial acceleration is the greatest; [1] as velocity increases, X increases, so resultant force of Y minus X decreases, and hence acceleration decreases; [1] the gradient of the graph steadily decreases with time; [1] after 8 s, X and Y are equal so resultant force on skydiver is 0 so her acceleration is 0; [1] skydiver travels at terminal velocity, and velocity–time graph is a horizontal line [1]

06.1 the resultant force is a single force [1] that has the same effect as all the original forces acting together [1]

06.2

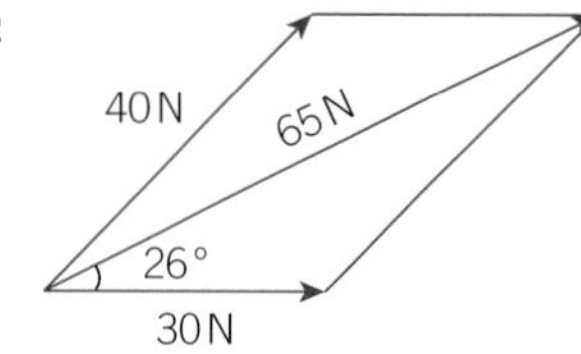

[3 N force and 4 N force drawn to scale – 1; angle of 90° between them – 1; completed parallelogram and resultant – 1; resultant of correct size and at correct angle (6.5 N at 26° to 3 N force) – 1]

P12.1 The nature of waves

1 they are compressed closer together than usual

2 transverse wave

P12.2 The properties of waves

1 10 m/s

2 waves per second

P12.3 Reflection and refraction

1 because waves change speed and wavelength when they cross a boundary between different materials

2 it does not change

P12.4 More about waves

1 an echo

2 sound waves are longitudinal waves, which always need a material to travel through

P12.5 Sound waves

1 decreases

2 20 Hz to 20 000 Hz

P12.6 The uses of ultrasound

1 20 000 Hz

2 ultrasound is not ionising

3 1500 m/s

P12.7 Seismic waves

1 solid

2 the focus

P12 Summary questions

1 a wave on a rope, for example [1]

2 $v = f\lambda$

$= 15 \times 0.45$ [1]

$= 6.75$ m/s [1]

3 the number of complete wave cycles passing a point each second [1]

4 by dipping the long edge of a ruler in and out of the water to create waves; [1] the faster the ruler is moved in and out of the water, the greater the frequency of the waves [1]

5 the vibration of particles in a longitudinal wave is parallel to the direction of energy transfer of the wave; [1] the vibration of particles in a transverse wave is perpendicular to the direction of energy transfer of the wave [1]

6 the closest point on the Earth's surface [1] to the earthquake's focus [1]

7 produce plane waves in a ripple tank; [1] place a piece of plastic under the water at an angle to the direction of travel of the wave fronts, so the depth of the water is less above the plastic; [1] refraction will take place as the waves move from the deeper to the shallower water [1]

8 it decreases [1]

9 a compression is a region in which the particles are closer together than normal; [1] a rarefaction is a region where the particles are further apart than normal [1]

10 3 kHz [1]

11 $f = \frac{v}{\lambda}$ [1]

$= \frac{0.75}{0.025}$ [1]

$= 30$ Hz

12 ultrasound is non-ionising; [1] so ultrasound is less likely than X-rays to damage the rapidly dividing cells of a developing baby; [1] ultrasound is reflected at boundaries between different types of tissue, [1] so it can be used for scanning soft tissues, whereas X-rays are mostly used for scanning hard tissues like bone [1]

P13.1 The electromagnetic spectrum

1 gamma rays

2 radio waves

P13.2 Light, infrared, microwaves and radio waves

1 e.g. for transmitting mobile phone calls, or for heating food

2 radio waves

P13.3 Communications

1 visible light and infrared

2 longer wavelength radio waves

P13.4 Ultraviolet waves, X-rays and gamma rays

1 lead absorbs X-rays, so lead aprons help to reduce the exposure of workers to ionising X-ray radiation

2 skin cancer

P13.5 X-rays in medicine

1 X-rays pass easily through soft tissue but are absorbed by bone, so cracks/breaks in the bone show up clearly on X-ray pictures

2 a measure of the damage done to a person by ionising radiation

P13 Summary questions

1 radio waves [1]

2 electromagnetic radiation containing a mixture of all the wavelengths in the visible part of the electromagnetic spectrum [1]

3 a very thin transparent fibre [1] that can be used to transmit communication signals by light and infrared radiation [1]

4 X-rays can be used to diagnose medical conditions [1] such as broken bones and dental decay; [1] they can also be used to destroy cancerous tumours near the surface of the body [1]

5 visible light, microwaves, radio waves [1]

6 infrared radiation is emitted by all objects; [1] the hotter the object, the more infrared radiation it emits; [1] infrared cameras detect the infrared emitted by objects, so the hotter the object the brighter it appears on the image [1]

7 can cause sunburn and skin cancer; [1] can cause damage to eyes [1]
8 gamma rays are used to kill harmful bacteria in food, [1] which may prevent food poisoning; [1] they are also used to sterilise surgical instruments; [1] the instruments are sealed in plastic and the gamma rays travel through the plastic and kill any bacteria on the instruments [1]
9 $f = \frac{v}{\lambda}$ [1]
$= \frac{3 \times 10^8}{3 \times 10^{-2}}$ [1]
$= 1 \times 10^{10}$ Hz [1]
10 optical fibres carry more information than radio wave and microwave transmissions [1] because they use visible light or infrared radiation, which have shorter wavelengths and so have greater energies; [1] optical fibres are more secure than radio wave and microwave transmissions [1] because they do not radiate electromagnetic energy, so transmissions cannot be intercepted [1]
11 the film in the badge changes colour if it is exposed to X-rays; [1] regular checking of the film badge means that the workers' exposure to ionising radiation can be monitored, [1] and preventative measures can be put in place if their exposure is too high [1]
12 the stomach is formed of soft tissue [1] so most X-rays will pass straight through it and not show up on an X-ray picture; [1] but the contrast medium is introduced into the body to absorb X-rays so the soft tissue will show up in the X-ray image [1]

P14.1 Reflection of light

1 a line drawn perpendicular to the mirror at the point where an incident light ray hits the mirror
2 the same size as the object

P14.2 Refraction of light

1 because a wave's speed changes when it crosses a boundary between two media
2 the angle between the refracted wave and the normal

P14.3 Light and colour

1 the range of wavelengths within the electromagnetic spectrum that can be detected by the human eye
2 an object that transmits all of the light incident on it; no light is absorbed at its surface

P14.4 Lenses

1 the point where incident rays of light which are parallel to the principal axis are brought to focus
2 place the object between the principal focus and the lens

P14.5 Using lenses

1 rays, drawn on a ray diagram, which come from a single point on an object to locate the corresponding point on the image
2 real; it is produced by real rays of light being focused onto the film or CCD

P14 Summary questions

1 virtual [1]
2 the angle between the incident ray and the normal [1]
3 it decreases [1]
4 converging [1]
5 a real image is formed when light rays from an object are focused onto a screen; [1] a virtual image is formed where rays of light appear to have come from, and cannot be focused onto a screen [1]
6 the light wave slows as it enters the glass; [1] unless the light wave is normal to the glass, it is always refracted towards the normal [1]
7 a line drawn perpendicular to a reflecting/refracting surface [1] at the point where an incident wave meets the surface [1]
8 a surface that absorbs all the light that is incident on it; [1] no light can pass all the way through it [1]
9 a white surface has no pigments [1] so it reflects light of all wavelengths either partially or totally; [1] white light is a mixture of all the visible wavelengths of light, so the surface appears white [1]
10 the distance from the centre of the lens [1] to the principal focus [1]
11 object height $= \frac{\text{image height}}{\text{magnification}}$
$= \frac{8.4}{2.4}$ [1]
$= 3.5$ cm [1]
12

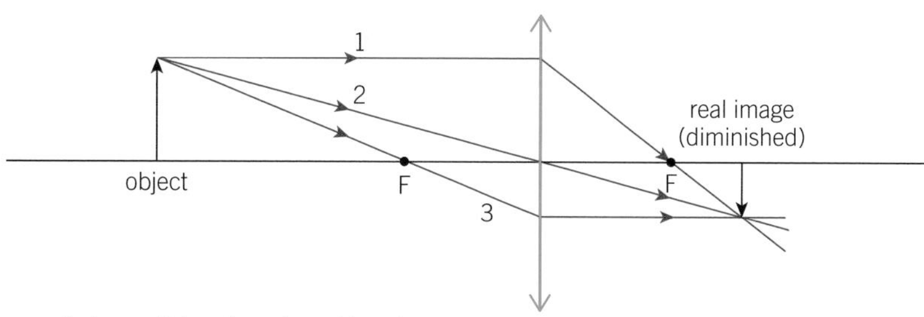

ray 1 is parallel to the axis and is refracted through F [1]
ray 2 passes straight through the centre of the lens [1]
ray 3 passes through F and is refracted parallel to the axis [1]

P15.1 Magnetic fields

1 for example: nickel, cobalt
2 steel

P15.2 Magnetic fields of electric currents

1 concentric circles centred on the wire
2 a long coil of insulated wire

P15.3 Electromagnets in devices

1 a solenoid in which the insulated wire is wrapped around an iron bar (the core); when a current is passed along the wire, a magnetic field is created around the wire which in turn magnetises the iron bar
2 a device containing an electromagnet that uses a small current in one circuit to turn on a larger current in a different circuit

P15.4 The motor effect

1 it stays the same
2 tesla, T

P15.5 The generator effect

1 the coil of wire must 'cut' the magnetic field lines to induce a p.d.
2 the direction of the induced p.d. is also reversed

P15.6 The alternating current generator

1 an alternating-current generator
2 a direct-current generator

P15.7 Transformers

1 a.c.
2 iron is easily magnetized and demagnetized

P15.8 Transformers in action

1 step-down
2 this increases the p.d. and decreases the current, which reduces energy losses in transmission

P15 Summary questions

1 it does not lose its magnetism easily [1]
2 the two magnets repel one another [1]
3 they are able to supply a magnetic field [1] which can be switched on and off [1]
4 increasing the number of turns in the solenoid; [1] or increasing the current through the solenoid [1]
5 the field lines curve around from the north pole of the bar magnet to the south pole [1]
6 when an unmagnetised magnetic material [1] is placed inside a magnetic field [1] and itself becomes magnetised [1]
7 the electromagnet is held over a pile of cans and switched on; [1] the steel cans will be attracted to the electromagnet and be lifted, but the aluminium cans will not; [1] the steel cans are moved to a separate place where the electromagnet is switched off, which releases them [1]
8 an alternator generates alternating current, [1] whilst a dynamo generates direct current; [1] a dynamo has a split-ring commutator whilst an alternator uses slip rings [1]
9 the left hand is held with the fingers and thumb all at right angles to each other; the first finger represents the direction of the magnetic field (north to south); [1] the second finger represents the direction of the current (positive to negative); [1] and the thumb points in the direction of the force [1]

10 when a magnet is moved into a coil of wire, the wire 'cuts' through the magnetic field lines [1] and a p.d. is induced across the ends of the coil; [1] if the coil is part of a complete circuit, a current will flow [1]

11 as alternating current in the primary coil of the transformer continually changes direction, the induced magnetic field in the core of the primary coil also continually changes direction; [1] for this to happen the core must be easily magnetized and demagnetized; [1] so soft iron is used for the core [1]

12 $V_s = \frac{V_p n_s}{n_p}$

$= \frac{2 \times 400}{100}$

$= 8\text{V}$

P16.1 Formation of the Solar System

1 fusion

2 protostar

P16.2 The life history of a star

1 iron

2 by a supernova explosion

P16.3 Planets, satellites and orbits

1 gravitational force

2 centripetal force

P16.4 The expanding universe

1 those furthest away from our own

2 it increases

P16.5 The beginning and the future of the universe

1 the theory suggesting that the universe started with a massive explosion from a very small initial point

2 it increases

P16 Summary questions

1 a repeating circular path of travel that one object in space takes around another one [1]

2 a region of space with such a strong gravitational field [1] that not even light can escape from it [1]

3 gravitational force [1]

4 wavelength decreases; [1] frequency increases [1]

5 it becomes a red supergiant [1]

6 protostar, main sequence star, red giant, white dwarf, black dwarf [3 marks – all correct; lose 1 mark for each incorrect stage]

7 the greater the relative speed of the galaxy, the greater the red-shift of light from that galaxy [1]

8 red-shift gives us evidence that the universe is expanding outwards in all directions; [1] one theory to explain this phenomenon is that if we imagine looking back in time, all matter would have come from a single point; [1] this is the Big Bang theory [1]

9 the inward force of gravity [1] is balanced by the outward force of the radiation from the core [1]

10 a resultant force [1] towards the centre of a circle [1]

11 light from a laboratory source on Earth contains dark spectral lines [1] that are caused by the absorption of specific wavelengths of light by certain elements in the source; [1] if the light from a distant galaxy is observed in the same way, scientists can see that that these dark lines are shifted towards the red end of the spectrum [1]

12 cosmic microwave background radiation was created as high-energy gamma radiation just after the Big Bang; [1] as the universe has expanded, this radiation has lost energy and increased to greater wavelengths, [1] and is now microwave radiation which comes from every direction in space [1]

Section 4 Practice questions

01.1 so the time between seeing the smoke and hearing the bang would be as long as possible; [1] to reduce the effect of reaction time [1]

01.2 so they could calculate a mean value for time; [1] to reduce the effect of random errors [1]

01.3 average time = 0.85 [1]

average speed = $\frac{280}{0.85}$ [1]

= 329 m/s [1]

02.1 volume of water in each can; [1] starting temperature of water in each can [1]

02.2 place cans on an insulating mat; [1] leave the cans for a longer period of time before measuring the temperature [1]

02.3 lower [1]

03.1 record number of turns of wire around the nail [1]; switch the current on and record the maximum number of paperclips, suspended end to end in a chain, which the electromagnet will hold; [1] add more turns of wire around the nail and repeat the procedure to determine how the maximum number of paperclips the electromagnet will hold has now changed [1]

03.2 the magnetic field will be stronger [1] so the electromagnet will pick up more paperclips, [1] provided that the number of turns of wire remains constant [1]

03.3 there will still be a magnetic field around the wire; [1] but it will be much weaker than the field with the nail in place; [1] so will not pick up many paperclips (if any) [1]

04.1 specular reflection occurs when light is incident on a smooth surface; [1] parallel rays of light are reflected in the same direction, [1] producing a clear image; [1] diffuse reflection occurs when light is incident on a rough surface; [1] parallel rays of light are scattered in different directions [1]

04.2 diagram shows reflection of one ray at the mirror, then entering the eye; [1] diagram shows reflection of second ray at the mirror then entering the eye; [1] correct path of virtual rays behind the mirror; [1] position of image correctly identified [1]

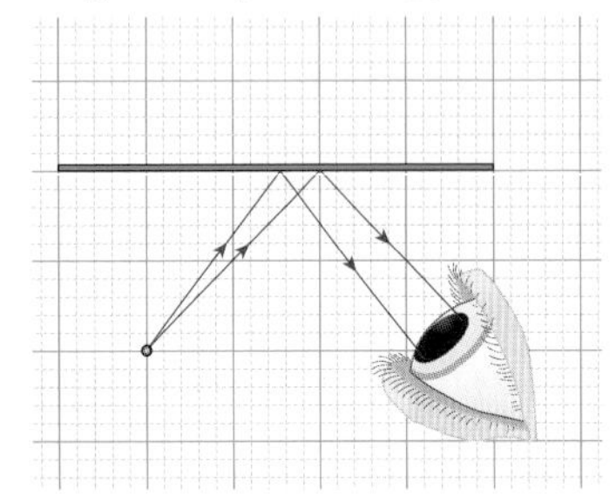

05.1 soft iron [1]

05.2 $n_s = \frac{n_p V_s}{V_p}$ [1]

$= \frac{11\,500 \times 20}{230}$ [1]

$= 1000$ [1]

05.3 the magnetic field lines of the primary coil must be 'cutting' the secondary coil in order to induce a p.d. in the secondary coil; [1] if the field is not alternating, no 'cutting' take place and hence no p.d. is induced across the secondary coil [1]

06.1 vibrations of the particles [1] are parallel to the direction of energy transfer of the wave; [1] so the wave travels through the air as series of compressions and rarefactions [1]

06.2 a coil that carries a current in a magnetic field experiences a force; [1] this is the motor effect; [1] the current from the amplifier varies, so the current in the coil varies, and so the force exerted on the coil varies; [1] the force moves the coil; [1] the coil moves the diaphragm; [1] the moving diaphragm sets up compressions and rarefactions in the surrounding air, which produces sound waves [1]

Practical questions

01.1 26.5 s [1]

01.2 $\frac{0.440}{26.5} \times 10 = 0.166$ m/s [2]

01.3 reaction time starting and stopping the stopwatch differed each time, or the difficulty in following the motion of the crest of a wave [1]

02.1 A – 0.5 A; [1] B – 3.0 A (2)

02.2 3.5 A [1]

02.3 the two lamps would be either both on or both off; they could not be switched individually [1]

02.4 in the parallel circuit, both lamps would be at 'normal' brightness because they each have a p.d. of 12 V across them; [1] because A's resistance is much higher than B's resistance, when they are in series most of the battery p.d. would be across A; [1] so A would be almost as bright as 'normal' but B would be much less bright [1]

03.1 153 mm [1]

03.2 308 mm [1]

03.3 extension = 308 – 153 = 155 mm [1]

03.4 $F = 5.0\text{ N} - 1.0\text{ N} = 4.0\text{ N}$ [1]

$k = \frac{F}{e} = \frac{4.0\text{ N}}{0.155\text{ m}}$ [1]

$= 26$ N/m [1]

03.5 repeat the test for four more different weights (e.g., 2.0 N, 3.0 N, 4.0 N, 6.0 N) and measure the extension of the spring for each weight; [1] **either** plot a graph of total weight on the y-axis against extension on the x-axis – the graph should be a straight line through the origin; [1] measure the gradient of the line – the spring constant is the gradient of the graph [1] **or** for each extension measurement, calculate the total weight + the extension to obtain a value for the spring constant; [1] calculate the mean value of the spring constant [1]

04.1 check the meters read zero (by opening the switch); [1] draw a table to record the ammeter and voltmeter readings; [1] adjust the variable resistor until the current is about 0.06 A; [1] measure and record the ammeter and voltmeter readings; [1] repeat the measurements for at least five more values of the current between 0 and 0.06 A at roughly equal spaced intervals [1]

04.2 

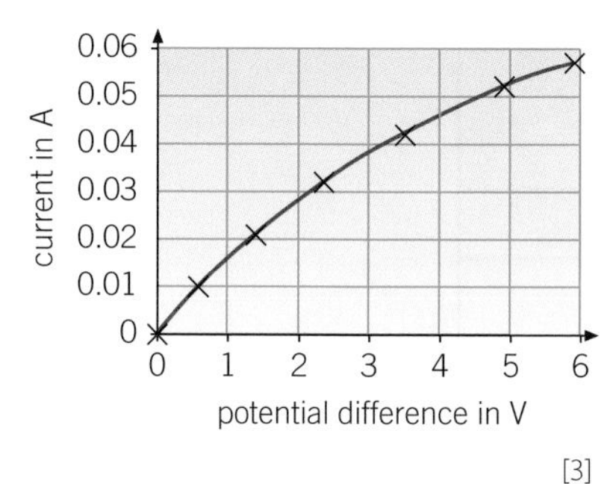

[3]

04.3 3.0V, 79Ω; 6.0V, 105Ω [2]

04.4 the current is larger at 6 V than at 3 V; the larger the current is, the hotter the lamp; [1] the resistance of the filament increases with increase of temperature; [1] this is because the atoms of the filament vibrate more and make it more difficult for conduction electrons to pass through [1]

04.5 $Q = It$

$= 0.06 \times (10 \times 60 \times 60)$ [1]

$= 2160\,C$ [1]

04.6 $E = QV$

$= 2160 \times 6$ [1]

$= 13\,000\,J$ (to 2 s.f.) [1]

Appendix 1: Physics equations

You should be able to remember and apply the following equations, using SI units, for your assessments.

Word equation	Symbol equation
weight = mass × gravitational field strength	$W = mg$
force applied to a spring = spring constant × extension	$F = ke$
acceleration = $\frac{\text{change in velocity}}{\text{time taken}}$	$a = \frac{\Delta v}{t}$
(H) momentum = mass × velocity	$p = mv$
gravitational potential energy = mass × gravitational field strength × height	$E_p = mgh$
power = $\frac{\text{work done}}{\text{time}}$	$P = \frac{W}{t}$
efficiency = $\frac{\text{useful power output}}{\text{total power input}}$	
charge flow = current × time	$Q = It$
power = potential difference × current	$P = VI$
energy transferred = power × time	$E = Pt$
density = $\frac{\text{mass}}{\text{volume}}$	$\rho = \frac{m}{V}$
work done = force × distance (along the line of action of the force)	$W = Fs$
distance travelled = speed × time	$s = vt$
resultant force = mass × acceleration	$F = ma$
kinetic energy = 0.5 × mass × $(\text{speed})^2$	$E_k = \frac{1}{2} m v^2$
power = $\frac{\text{energy transferred}}{\text{time}}$	$P = \frac{E}{t}$
efficiency = $\frac{\text{useful output energy transfer}}{\text{total input energy transfer}}$	
wave speed = frequency × wavelength	$v = f\lambda$
potential difference = current × resistance	$V = IR$
power = current^2 × resistance	$P = I^2 R$
energy transferred = charge flow × potential difference	$E = QV$
pressure = $\frac{\text{force normal to a surface}}{\text{area of that surface}}$	$p = \frac{F}{A}$
moment of a force = force × distance (normal to direction of force)	$M = Fd$

You should be able to select and apply the following equations from the Physics equation sheet.

Word equation	Symbol equation
(final velocity)^2 – (initial velocity)^2 = 2 × acceleration × distance	$v^2 - u^2 = 2as$
elastic potential energy = 0.5 × spring constant × extension^2	$E_e = \frac{1}{2}ke^2$
period = $\frac{1}{\text{frequency}}$	
(H) force on a conductor (at right angles to a magnetic field) carrying a current = magnetic flux density × current × length	$F = BIl$
change in thermal energy = mass × specific heat capacity × temperature change	$\Delta E = mc\Delta\theta$
thermal energy for a change of state = mass × specific latent heat	$E = mL$
(H) potential difference across primary coil × current in primary coil = potential difference across secondary coil × current in secondary coil	$V_sI_s = V_pI_p$
(H) pressure due to a column of liquid = height of column × density of liquid × gravitational field strength	$p = h\rho g$
(H) $\frac{\text{potential difference across primary coil}}{\text{potential difference across secondary coil}} = \frac{\text{number of turns in primary coil}}{\text{number of turns in secondary coil}}$	$\frac{V_p}{V_s} = \frac{n_p}{n_s}$
For gases: pressure × volume = constant	pV = constant
(H) force = $\frac{\text{change in momentum}}{\text{time taken}}$	$F = \frac{m\Delta v}{\Delta t}$
magnification = $\frac{\text{image height}}{\text{object height}}$	

Great Clarendon Street, Oxford, OX2 6DP, United Kingdom

Oxford University Press is a department of the University of Oxford. It furthers the University's objective of excellence in research, scholarship, and education by publishing worldwide. Oxford is a registered trade mark of Oxford University Press in the UK and in certain other countries

First published in 2017

British Library Cataloguing in Publication Data
Data available

978-0-19-835942-5

10 9 8 7 6 5

Printed in Great Britain by Bell and Bain Ltd, Glasgow

With thanks to Jim Breithaupt for his contribution to the Practicals support section.

Acknowledgements

COVER: Eric James Azure / Offset
Header Photo: Niyazz/Shutterstock;
p9: Kzenon/Shutterstock; **p16(L)**: Iakov Filimonov/Shutterstock; **p16(R)**: Zeljko Radojko/Shutterstock; **p18**: Julof90/Istockphoto; **p23**: David Giral Photography/Getty Images; **p33**: Ulkastudio/Shutterstock; **p34**: Cordelia Molloy/Science Photo Library; **p36**: OUP; **p74**: Conrado/Shutterstock; **p75**: Thieury/Shutterstock; **p90**: Gagliardiimages/Shutterstock; **p96**: Martyn F. Chillmaid/Science Photo Library; **p97(T)**: Wang Song/ Shutterstock; **p97(B)**: Cnri/Science Photo Library; **p114**: Dedek/ Shutterstock; **p117**: Walter Pacholka, Astropics/Science Photo Library;

Artwork by Q2A Media